The nurse and the "crazy"

Leonardo Massi

The nurse and "the crazy"

Original Title: L'infermiere e il "pazzo"

Author: Leonardo Massi

Copyright © 2013 Lulu

Cover Design © 2013 Leonardo Massi

Cover Image by Leonardo Massi

Translated in English by Georgiana Gherghina & Leonardo Massi

ISBN: 978-1-304-65345-1

All rights reserved (included the cover image). No part of this book shall be reproduced, stored in a retrieval system, or transmitted by any means, electronic, mechanical, photocopying, recording, or otherwise, without written permission from the author. No patent liability is assumed with respect to the use of the information contained herein. Although every precaution has been taken in the preparation of this book, the publisher and author assume no responsibility for errors or omissions. Nor is any liability assumed for damages resulting from the use of the information contained herein

Enquiries concerning reproduction outside the private reading should be sent to the author, as well as any other inquiries, at the following email:

leonardomassi@ymail.com

Warning and Disclaimer

Every effort has been made to make this book as complete and as as accurate as possible, but no warranty or fitness is implied. The information provided is on an "as is" basis. The author and the publisher shall have neither liability nor responsibility to any person or entity with respect to any loss or damages arising from the information contained in this book.

ABSTRACT

The essay leads the reader in a colloquial and informal way, but at the same time yet rigorous, through the meanders of the relationship between "normal/crazy". On one side the patient, on the other the nurse; on the background the same social context by which both of them are being evaluated. The analysis of the madness through a social and multidisciplinary view. A light dancing through different disciplines in order to better taste the strong flavor of the madness in the patient and its bitter-sweet aftertaste in the nurse.

The true mystery of the world is the visible not the invisible ...

Lord Henry from *The Picture of Dorian Gray,* by Oscar Wilde

Life goes grip, even with dentures! (Life must be bitten, even with dentures!)

Anonymous retired old lady from Milan

CONTENTS / INDEX

Acknowledgement	*7*
Anthropological Introduction	*9-30*
Interview to Gaudio	*31-63*
Failed Interviews	*64-70*
Apology (not required) of the nurse	*71-73*
The Other Human Focal Point /Focus	*74-97*
Bibliography	*98-102*

ACKNOWLEDGEMENTS

I cannot begin without thanking people who do not even know how deeply I feel indebted to them. A multi-disciplinary approach to the topic makes me thank many people, most of all to two of them: the nurse Claudio and the expert Simone Magi. I would also like to express my gratitude to prof. Desideri and to prof. Nocentini for giving me a brilliant example of academic research; to prof. Brilli for the care with which he devoted himself to the teaching of mathematics, and to prof. Cincilla for his lectures (and the teaching methods used) at the time of the high school. A special thanks to all the extended family and to the two newcomers: Enrico Bertoni and Nicoleta Gherghina. A special thanks to Veronica Massi, Francesca Massi, Andrea Tanci, and last but not the least, Georgiana Gherghina, for the unfailing support. I would like to thank all the students to whom I have had the pleasure to teach in the past two years (a profession that initially I did not want to undertake), the few students of Besta, the many of Mazzotti and the crepuscular students of the evening course of Sirius. I'd love to name them all, one by one, but because of space problems I postpone this to a different context (which for sure will not be the assigning of the final grades ...) and I confine myself in thanking them collectively.

Finally allow me to thank all those people who I can no longer touch with these hands.

INTRODUCTION

We want to present here and introduce two human conditions. The nurse and the "crazy". Each of them is presented using different modalities, as are different their two social contexts. For the nurse is used an interview in anthropological style, for the "crazy" a literary "impressionist painting" (an allusion that you are going to understand in the chapter dedicated to him), which certainly is neither less "real" nor less serious than the first one.

In truth one could read this introduction, and understand it better, after reading the interview itself. So one might then skip it and read it later, as you could do with a hypertext, but here, however, instead of a command to translate digitally our will there is only the slight move of the hand on the edges of these pages. Browse them because they are true, in every sense. At this point (rightly) one could object questioning what is the advantage of putting this introduction at the beginning. To be honest I put it at the beginning just to pay tribute to that sense of security that many of us have in following what traditionally was always seen to be done "so". And there is nothing more certain and more traditional than an introduction placed at the beginning.

The piece of writing focuses on an interview of a nurse of a mental health center, being an anthropological interview therefore it requires some initial gloss. Let's talk a little about anthropology. According to the definition of Devoto and Olii present in the LeMonnier vocabulary, anthropology is "the science that studies the human types and aspects, especially from a morphological, physiological, psychological points of view". This definition fits an anthropological "field" study, as it might have been the one carried out at the time by Malinowski. It fits well to the prolific definition of Tylor too, which yet I reckon does not fully reflect the contemporary anthropology studies. [The definition of Tylor : The culture or the civilization , understood in its wide ethnographic sense , is that complex whole which includes knowledge, beliefs, art , ethics , law, custom and any other capabilities and habits acquired by man as member of a society. Tylor 1970 (1871) p.1] The latter ones (these contemporary anthropological studies) are in fact less exotic and in some ways maybe even less descriptive and demonstrative (ostensive) than the "classic" studies of anthropology, but in other aspects are much more introspective. Anyway, the previous definition is intended and fits perfectly to the kind of anthropology which precisely is now definable as " classic ", in other words that anthropology conceived and applied until some decades ago. Regarding the first part of the definition of Devoto and Olii, I do not know if anthropology can be defined as a science, Popper would object, and after all, the word " science" is subject to be interpreted in a myriad of different ways and even more aspects; as to the second part of that definition I repeat what I have just said: that is attributable to the classic anthropology studies rather than to the modern anthropological studies. I would prefer to define anthropology not so much as an analysis of the human cultural forms, but rather as a quite precise analysis on the man by the man.

This definition is supported by the recent developments of this discipline. I seem to notice that in past years the anthropological studies do not point exclusively to the exotic, whose field of action now is indeed very small compared to the past (inter alia from many of those "exotic" areas the scholar of anthropology has been banned because of political reasons, especially in the early stages subsequent to decolonization, but they rather aim to what we could call an "interior exotic". This explains the great attention that since the second world war anthropologists gave to the Mediterranean sea [In reference to this , see the various essays published in *Anthropology of the Mediterranean* , edited by D. Albera -A . C. Blok - Bomberger , Milan, 2007]. An "exotic" east and south were lost, and it was tried to find that "exotic areas" in places accessible to the scholars. He who seeks shall find (Italian proverb) (*seeks and ye shall find*, from the Bible). This has led us to cast new light on the anthropology. No longer as just the study of the different, no longer as solely the reconstruction of the human history, or on the contrary as a simple end of a cultural relativism in which everything is justifiable, but rather as the proper study of the man carried out by the man on the man. It is also true that we are no longer within a linear evolutionary vision of the human and natural history. But it is also true that this kind of conception of history is not a concept to jettison if we start from the awareness of the connections linking the various cultures during their genesis and of the total nonexistence and groundlessness of "pure" cultures. (See the interesting position of Fabietti, as we can see in Fabietti 1998). On the other hand if you uncritically throw away the idea of a history in evolutionary sense, what remains in the general mentality is a " cultural relativism" certainly no less prolific of problems. In relation to the cultural relativism is good not to be ambiguous. I am more inclined to accept from a cognitive point of view a moderate epistemological relativism, which leads me to be critical of an uncritical, pardon the pun, a cultural relativism, under which surreptitiously lies an absolutism in which all points of view are put on the same plane. We cannot turn a blind eye unless being hypocrites, and then we cannot fail to recognize a certain our propensity, more or less veiled, more or less conscious, to consider the history in an evolutionary sense. And if this is related to our current cognitive mechanisms (and I stress current !) (And since I started this topic I underline that by using this terminology I refer to a world and therefore to a way of thinking essentially mechanistic and evolutionary; my actual world), or if it is linked to a relevance to the world outside of our minds, or more, I do not know and nobody does either. I would be tempted to add a "for now", but in doing so I would fall in one of the prospects earlier criticized. On the other hand, the intellectual honesty is not in the avoidance of a question, but in recognizing it and in setting limits for it. Herein that is easy for me since the task of this brief essay is not to solve the problems highlighted.

Popper, unlike the linguists Devoto and Olii, has never considered Anthropology, as well as Sociology and Marxism, a real science, but, in opposition to the previous, he considered Physics, Astronomy, Biology etc. as real sciences. It is to be noted that since the 60's the criticism to the scientific inquiry as it is described both by him and by a "naïve" positivism which develops from Bacon forward, lead more than to the weakening of the concept of scientific research (position hailed by some and considered by me grotesque) to exactly the opposite outcome, that is a revaluation of what was seen at a certain time by Popper, and not only by him, like not scientific. [In reference to the "naïve" positivism: that "naive" term does not have an evaluation function but only a descriptive one. It is clear that that initial positivism was anything but naïve. We consider in this manner the term "naïve" only retrospectively in order to distinguish it from the subsequent and various philosophical positions that the positivism has acquired over time. In this classification I have followed what was suggested by Ladyman (Ladyman, 2009 [2002]). In reference to this see the circles of Vienna and Berlin of the early '900.]

Like my aforementioned definition of anthropology, the one given by Fabietti, Matera and Manighetti [Fabietti, Malighetti, Matera, 2002, p.9] is also very intriguing, and in some ways is more relevant than mine being more circumscribed: "Anthropology is therefore , in a certain sense, the heritage of the world." I like this definition. A posteriori it seems to me the one that best embodies the anthropological interview conducted by me (the one you are going to read). In fact, from the words of the interviewee transpires some semblance or form of a legacy that the interviewee gives and directs not towards his present and future colleagues, but to any present and hypothetically future reader. This perspective did not emerge in obedience to my will. I only tried to make the interviewee speak about what he considered the most appropriate for the topics we were dealing with. I emphasize that this perspective has emerged from the words of the respondent. I do not know if he was aware of it or not, but as he proceeded to speak, both in some of his apologetic hints and in other hints of different nature, as we shall see, the interviewee has given the impression of wanting to leave an inheritance, a gift to be guarded by some community or future subject.

Maybe my first academic curriculum as a historian leads me to consider that a historical research intended to investigate the deep causes of an event will necessarily end up in anthropology. [The connection between anthropology and economics, which is in my opinion at the base of any event, is complex and not suitable to be addressed here.] To confirm this, you can take an ancient example and a modern one. The ancient example. Tacitus writes between the end of the first century and the beginning of the second century AD, its historiography goal is the search of the root causes that led to the crisis of the Roman Empire. In truth, his investigation had started in a time when the re-establishment of freedom, beginning from Nerva, gave to the historians the possibility of a free historiographical analysis, which in the case of Tacitus was addressed in an investigation into the "positive" present compared to the just ended "negative" past. [The type of this research already informs us about the relativity of this " freedom".] That was the goal of the *Historiae*, which then had drifted into the later *Annales* in an ever more distant and deeper exploration of the past and of the human soul. And therefore drifted from a survey designed to describe the transition from " bad " to "good" into a research in the opposite direction: from good to bad. The question is: how could it happened? His pushing forward , his investigation, led him not only to go more and more back in time, contrary to what he had previously planned, but to investigate even more the " human" reasons immersed in the specific investigated contexts. From that specific context, from those specific people, to human being, to human way of acting. Reasons not even too hidden in his other two historiography works: *Agricola* and *Germania*. And bear in mind that we are talking about a historian who lived between the first and the second century AD. The other example that I propose, the modern one , is embodied by Habsbown, a contemporary historian. In his great work (*The Short Century*) Habsbown flows into anthropology whenever the wealth of data allows to investigate the social changes which shook the foundations of Western society. For example (E.g.), investigating the reasons for the occurrence of certain social unrest starting from the 50s of the last century (especially the 60s), the answers by him provided have a particular anthropological characterization. Unfortunately, the anthropology as it is today proposed (and perhaps also, unfortunately, as it is currently understood) in most of those same rooms that can accommodate academic classes converging to it , has become a too much, far too much , reductive concept (an understatement; simplistic concept). A desk Anthropology that produces as a mere as futile rhetoric exercise has very little to teach those people seeking interpretive models of the "real".

Nowadays anthropology must not aim only to this futile rhetoric exercise. Far from it. I think the society, our society, is permeated by a strong social need for anthropological expertise (anthropological skills). And I also believe that anthropology can provide enlightening keys to understanding the surrounding world. See for example the work of D. Harvey, *The crisis of modernity*. At the same time I believe that many university courses related to anthropology are almost completely useless (I leave an "almost" for the confidence I place in some background noise that presumably should remain in the unconscious). Nowadays we tend to explain every phenomenon, including, of course, those having an anthropological matrix, by means of the economy (which interprets a powerful perspective of the "real" world, extremely fascinating). But often we could have even more comprehensive readings if we analyzed these same phenomena with an opposite trend, that means describing the economy through anthropology. Indeed I think that anthropology and economics are two disciplines inextricably linked. Actually as all the disciplines. Anthropology and economics speak the same language and describe the same world, but the first one is more and more dominated by the latter one for its inability to look at " today " with the proper standards of "today", with the appropriate standards of the "right here and right now". Usually we use boxes and containers, so let me respond with boxes and containers. In practice, in general, an economist or an engineer who has the anthropological bases is much more useful than an anthropologist who has economics or engineering bases. And this happens in a society where in my opinion, as I said, there is a profound need for widespread anthropological skills. Most of anthropology at times seems to be limited to the defense of certain social categories or those that seem to be the most vulnerable groups of people, sometimes it seems addressed only to the classification and crushing of mankind into groups well separated in watertight sealed compartments, through the ethnicization of the same human communities. An ethnicization which is often most prolific of consequences than full of assumptions [Fabietti, 1998]. The majority of modern anthropology does not produce an active, aimed, voice in society, producing instead a light and soft lament, but at the same time quiet like a person who releases it from a position of privileged respect to the object of his theatrical defense. Leads to regretting a past time or illustrating a world that would be more " fair", into denouncing the contemporary discrimination and injustice from a beautiful little room. All with the same vital force of a sloth. Often lacks the understanding of the " anthropological " reasons of those changes, it lacks a strong philosophical framework through which to understand the society and the individual.

One clings to the anthropology not to investigate " the man ", but to keep afloat a reassuring model of " man", or more often even, unfortunately, a specific man: the anthropologist himself.

Precondition: the aim of this brief introduction is to present the interview touching briefly on the background issues that a broader and a more ambitious research than mine should address. The topic of the interview is about the work of the nurse and therefore his relationship with the patients commonly known as "crazy/insane people". Without entering the gnoseology (theory of knowledge) and avoiding the familiar debates related to linguistic categories in connection with our cognitive apparatus, it is worth noting that in order to speaking of " crazy/insane people", that means speaking of this specific group of patients subject to special medical care, one should have in mind a model of them, a stereotype. We should therefore investigate our own cognitive apparatus, and in this case the study should start with a long introspective dissertation on such matters, and as a consequence directed mainly towards the interviewer. Remaining faithful to the original assumption , and so without entering in the epistemology of the investigation , I will just briefly run through the logic and methods used for the approach to the categorization of patients applied in this interview. In this regard it is appropriate to mention the fuzzy logic theory, as a logic suitable to explain our category of "crazy " from which sprouts the stereotype we refer to when we talk about " crazy people". Fuzzy logic is connected to a theory of knowledge, strongly connected to our technology sector, as illustrated by Boncinelli [Boncinelli 2006]. I want to stress that this is valid both if the technological environment is understood as a product and if it is understood as cause of our human society. In other words what is stated above works both for a determinism that makes deriving the technical means from a social tool, that means it comes from the society that gives birth to it [see Offner 1996 I refer in particular to that beautiful expression, as hermetic as deep, that quote at hand is: "the technical mean is always a social tool"] and it also works for a determinism of the opposite direction, for which you may refer to the common sense. Cause and effect often, very often, intertwine each other and the precedence of one over the other seems to be due to the perspective from which we look at the same events . You could compare the cause and the effect to the two sides of the same coin. However talking of prospects instead of sides of the same coin satisfies me more: the prospects could be more than two, even though we do not know it or we cannot know it. On the other hand I do not believe in the principle of cause and effect except as a functional interpretation of the reality, from a human point of view.

The debate nowadays places at the core of the reflection a mind that works with parallel modes, cognitive networks, implicit and ductile models, prediction modes, alterations/rearrangements/reshuffling subsequently acquired to experiences. Thus the most fruitful theoretical references seem to be those of the connectionism (more connections), of the proto typicality, of the "politeticità"/polytheticity (literally: diagnosis that contemplates more criteria), the fuzzy concepts.] (Breda 2000, page 47)

According to recent theories developed in mathematics area (about the so-called fuzzy sets) and explicitly starting with the sixties (Zadeh , 1965, 1968), the fuzzy logic would be a kind of filter used by our brain, which starting from the identification of a prototype proceeds fading its attributions and assigning them to other elements of the surrounding world. This happens according to a scale of values, conventionally from 0 to 1. [The fact that this theory has not an absolute value but is, like any other theory, connected to our current historical period and to our technological development, it seems to me quite clear and not subjected to further explanation that would be here only superfluous. On the objection that what I have just written would be internally contradictory, because denying in an absolute sense that there isn't an absolute statement it would fall in the paradox, I would answer with the same words with which Nietzsche replied on the pages of *The Birth of Tragedy* to those who accused him of contradicting himself : " And if anyone objects that what I say is contradictory, and if in what I say there is contradiction. Well, so much the better !" The problem is vast, but unlike Nietzsche, having no esoteric intention, I can only recognize that, strictly following a logic of cause and effect, the contradiction still remains, but even if we seem to work by the cause-effect logic, we are not allowed, at least at the present, to know if there may be other working (functional) logics too.

That means we cannot even deny it.] Such logical process is used in anthropological research too. Therefore it is clear that in this context, the identification of a prototype is essential. It is the heart (center) of a category from which stretches (sprawls) a periphery more and more nuanced and distant. The point is that in the anthropological adaptations of fuzzy logic, the "heart (center)" is itself subjected to a nuances of meaning. Piasere himself (Piasere 1998) speaks about prototype both as the "perfect representative" of the reference category (such as may be considered the holy family for Catholics) and as a representative of the class that has the largest number of characteristic features, but not necessarily all. It is clear that this dual definition supports Popper's theory about considering anthropology as non- scientific, and therefore it is contrary to the same intent of Piasere. As a matter of fact Piasere leans against thesis that already in their dual definition create theories which will always find solutions in themselves, solutions that are self-validating. So they may not be subjected to the well known (empirical) falsification. Now as mentioned earlier, the confidence that Popper has in this principle (the falsification) to demarcate what is science from what is not, is entirely misplaced, as the subsequent philosophers of science Hanson and Kuhn (Kuhn 1969 [1962]) and also Feyerabend have shown. But it is also quite evident that Piasere should have a clearer analytical line, that although would be subject to more critical considerations, at the same time would be not only bolder but also more prolific of further developments.

Following this fuzzy logic, regarding the matter of the interview, the prototype that I was looking for, related to the category of the "insane people" would have been given to me by the expert nurse. At least according to my expectations. But when the respondent was induced to talk about his patients on my request, he always avoided some kind of definition of "crazy people". At best, he reinforced the varieties of "crazy people" enunciating several other sub-categories by saying that there are people with schizophrenia, with obsessive neurosis or hysteria etc., but he never gave me a comprehensive definition neither prototype nor stereotype. The fall into an analytical taxonomy (classification) more and more complex concerning those that for me, ignorant of the matter, were patients of the same category, slips away/escapes from a definition of the prototype. It is moreover clear and expected that the more you master a subject, the more the survey (research) breaks up that field of action. In conclusion, I have never been set before some explicit prototype by the interviewee. On the other hand I have to assume that the interviewee must have had some prototype in his mind, at least on a subconscious level, considering that for him his patients were those to whom he had to be able "to make them socialize by themselves" (you will understand it hereinafter). That is part of the ultimate goal which is to try to give his patients the means to build themselves a "normal" life. In his job, the goal of the interviewee was the same for each of his patients, always the same. The patients could be classified under different dysfunctions /pathologies, but they all had to be guided to a "normal" relationship with the other people. Perhaps this is that core, that heart, of the fuzzy theory searched by me and from which branch off all those nuances connected in this case not only to single diseases/pathologies but also to "degrees" more or less intense of the same individual pathology.

Speaking of the mentally ill, the noun and the adjective "normal" appear frequently. It was a fortunate fact, but perhaps also the natural outfall of our chat, that at the end of the interview we went back to discuss about the relation of his patients with the "normal" society, the one that is outside the mental health institutions. From that analysis the respondent started a speech full of pauses landed in a " but in the end we all are ", you will understand the allusion later. I am an ignorant of the matter both from a clinical diagnosis point of view and from any point of view that is based on experience and direct contact with people suffering from various problems of a psychic nature. But in my ignorance I always thought that normality is to be considered a chimera, fruit of the statistics, modern and rationalizing/disciplining science of the real (or the outside world) par excellence. The normal man is the result of the average between individuals different from each other, which then by definition are not normal. In this regard, I would point out that an "insane" individual becomes a patient only after passing through a social assessment of the society in which he lives. I would like to cite a passage from an essay by Schrippa, which as an European, and especially as an Italian, has gone for a few years to Ethiopia to study those which according to him "can be defined as the processes of health and disease".

Lavorare sui processi di salute e malattia significa interrogarsi su come gli individui e le società fanno fronte, concettualizzano, e prevengono ciò che viene in qualche modo visto come una minaccia per la salute, e dunque per gli individui e la società stessa. La salute, e ancor più la malattia, costituisce per ogni società un problema: la minaccia alla salute, cioè la malattia (in questa accezione vengono considerati anche gli "insani di mente"), esige che ogni volta che essa si presenti, si possa ricercare una spiegazione, una interpretazione; occorre cioè, per riprendere la felice espressione di Marc Augé e Claudine Herzlich (Augé-Herzlich 1986), che si trovi "il senso del male" affinché gli uomini possano sperare di controllarlo e, possibilmente, di vincerlo. (Schrippa)

[Working on the health and disease processes means to wonder about how individuals and societies cope, conceptualize, and prevent what is somehow seen as a threat to health, and therefore for individuals and society itself. Health, and even more the disease, form a problem for any society: the threat to the health, i.e. the disease (in this meaning are also considered the "insane"), requires that whenever it occurs, an explanation, an interpretation can be searched for; what is required, to resume the happy expression of Marc Augé and Claudine Herzlich (Augé-Herzlich 1986), is to find out "the sense of evil" so that men can hope to control it and, possibly, to win over it.] (Schrippa)

The interviewee, as it will be seen, points out that many of his patients, or at least all those who he directly quotes, come out from " social contexts poor both intellectually and economically". Beyond the legitimate economic considerations that could be done on the issue of inequality against the illness, here what I want to emphasize is that in addition to a purely economic issue there is a social one too. In fact the risk, or the responsibility, that hangs not only on the medical staff but also on the political staff regarding the management and the decisions concerning the various health institutions (and not only), the risk (I said) inherent in the process of medicalization is the mystification.

[In fact the interviewee often refers to the policy as an element which determines the development, or not, of certain clinical environments in comparison to others. This will be seen into detail when that subject will be touched during the interview. But let me say here, that what for the interviewed nurse is a fundamental guideline given by the policy to the " medicalization " and to the medical research, it seems to me only a pale small mirror that reflects what is the aim towards which the society decides to go. Priorities change with the changing of society, even in the medical sense. And one example is precisely the evolution of the structures and of the interpretation of the madhouse following the new concepts of " care " occurred over time.] That is making exclusively natural and exclusively registered in the biological and physiological processes of a single individual person, what is instead also the physical registration of social relationships; not everything is inscribed in our biological and physiological processes. The risks of the existence are not the same for everyone. That contemporary denomination of each physical or nervous illness that we have learned to name "stress", that dark suffering so often baptized in a medical sense with "depression", or what that even in Virgil (*The Aeneid*) and in Horace turns out to be a numbness that distances us from life and that makes us lose importance to everything surrounding us, are all individual and not contagious "diseases", as is the "madness". Even Philippe Pinel (1745-1826), physician and psychiatrist who practiced with great honor during the French Revolution and the Napoleonic Empire (historical moment that gave rise to a new concept of Europe and of Man, and in which previously matured new philosophical and social instances materialized) and who was the first in the Modern Age to distinguish the mentally ill people from the homeless people and the marginalized people to whom they were formerly associated, recognized the " melancholy " as a disease of the body, one of the five possible mental illnesses. And it was according to this "social recognition" of the "sick persons" that were created the modern asylums/madhouses, places where individuals deemed by society as incapable of mastering one's instincts were recovered to be constantly monitored by a doctor who followed the evolution of the disease. And of course in this way they were removed from "normal" society. The melancholy, the depression, the alienation are "diseases" that affect everyone, without distinction. The inequality in front of the surrounding (external) world registers strongly in our individual history through our body, and then(accordingly) through the registration of the body in the social order. Individual action and the body itself are interpreted within the society. I believe that whether we speak of community or society, the familiar question gemeinschaft / geselschaft inaugurated by Tonnies and deepened by Durkheim (Durkheim 1962 [1893]), this interpretation always assumes the translation of an assessment to which the individual is subject. The society interprets the individual and then places him right inside. It is almost like in the Verga's fiction where the "chorus" of the society interprets the action of its own people, of its own voices. The following quote refers to a context broader than that one treated by me, but incorporating our argument too, and being heuristically relevant, I like to conclude so:

Assistiamo infatti a una continua disciplina di una serie di atteggiamenti della vita quotidiana che in tal modo vengono fatti rientrare nel quadro di una normativizzazione che discende da un sapere medico che si fa disciplina del corpo e del comportamento. La possibilità di intervenire medicalizzando, di dare regole e comportamenti di vita, di costruire un soggetto attraverso tali regole, avviene non quando insorge la patologia ma nel momento in cui l'obiettivo è il mantenimento degli organismi in un certo stato che viene definito come "salute".

We are witnessing a continuous regulation of a set of attitudes of daily life which are thereby brought within the framework of a normativization that descends from a medical knowledge that makes itself discipline of the body and of the behavior. The possibility of intervening by medicalizing, of giving rules and behaviors of life, of building a subject through these rules, does not happen when the disease occurs but it happens when the goal is the maintenance of the organisms in a certain state that is defined as "health."

Before going into the interview I would like to say a few words about a fairly central point about the daily work done by the interviewee and pointed out by himself. In his job it is of vital importance to build a relationship of trust with not only the patient but also with the family of the same patient. At the macro level you can define "the trust" as:

aspetto strutturale delle organizzazioni societarie in quanto presupposto e condizione dell'interazione. (Fasulo 2002, pag. 51)

structural aspect of the social organizations as a prerequisite and condition of the interaction. (Fasulo, 2002, page 51.)

This definition applies at the micro level too, i.e. the interaction between two men. The relationship of trust that establishes between two persons can be considered in a fundamentally competitive way as in Goffman (Goffman 1988 [1967]) or in a more cooperative perspective way as in Luhmann (Luhmann, 2002 [1968]). Fasulo is keen to stress that anyway both anthropologists consider the trust a by-product of the action. It is obvious that different relationships between human beings are the basis of different constructions of trust. This gets across clearly in the interview. The interviewee speaks consciously of two different "constructions" of confidence/trust: the one with the patient and the other one with his family. These two interlocutors are not the same thing and as a consequence the relationship of trust must be built on different bases. The Goffman's competitive vision, which puts the deference and the demeanor at the base of the mutual attribution of trust between two persons, fits perfectly to the relationship that the respondent says he wants to build with the patient's family. In this context, to gain their trust, he seeks to meet the needs of the patient's family, but in order to earn a productive trust for his working context, he emphasizes the fact that one should not go too far with the concessions "otherwise you will lose bargaining power". Instead, we can consider the more cooperative vision of Luhmann to refer to the relationship that the interviewee (respondent) seeks to build with the patient. It is clear, however, that in him there is the awareness that this trust is not based on an equal distribution of the information: he is the one that helps the patient, the patient is the one who needs help from a social point of view (so regardless of the actual request for help by the individual-patient). As a result, according to Fasulo in a pertinent quote from McLeod (McLeod , 2008), we can say that in this case:

(il) dare fiducia implica una posizione sistematicamente ottimistica riguardo all'altro, ma che non investe la totalità della persona.

giving confidence implies a systematically optimistic position about the other, but that does not invest the totality of the person. (Fasulo 2002)

The difficulty of the nurse specialized in this type of relationship with such patients is due to the fact that the cooperative relationship proposed by Luhmann, foundation of the trust between two individuals, clashes in sharp contrast with the control of the other. The exercise of control, that the nurse must have during situations with his patient, would tend to defuse the process of trust. It is this light dance between these two different contexts to make very difficult the job of a nurse of a mental health center. On one hand building a relationship of trust with the patient, on the other hand having the control over him.

INTERVIEW TO GAUDIO

[In the transcript of the interview I try to leave the dialectal forms used by the interviewee and myself, without any phonetic expedient. The graphemes used will be those of the English (Latin) alphabet. Since the interview was conducted in Italian, I will try to make it as faithfully as possible to the original "sounds" in order not to lose that concept inseparably intertwined between the sound and the meaning of the words. But many times I'll have to concentrate more in making the meaning rather than the sound, for this fact I apologize. We are in Italy, to be more precise in Città di Castello, a town of about 40,000 inhabitants, located on the green hills of Umbria, roughly halfway between Florence and Rome.

The interview is reported below and only occasionally interrupted by brief comments. As you will see Gaudio is very clear and does not need further explanation. The phrases in parentheses are not part of the direct speech of the interviewee, but they are only my notes. In the next chapter I linger mostly on making a parallel between the interview given to me by Gaudio and those not given to me by others. Clearly the reason for this subsequent chapter (linked to this one) is not a revenge for what I asked and did not get, but because I think that the untold, entirely legitimate and in some ways understandable, has not much less value than the told (talking). Every comment made by me is just a figment of my personal considerations, not even supported by previous experience in the field.]

When I went to the "pink house" to ask for a nurse to interview, I rang the bell and waited outside the glass door of the building where its offices are located. The pink house is in fact right in front of it. The classic house that such a word evokes in one's mind: square, roof in the genre of the most common ones, white windows arranged symmetrically. Only the dimensions distance it slightly from the banality, they are slightly bigger than for a normal house. After all a signboard placed in front of the building, outside the fenced yard, names this building "Villa Rosa" and not "pink house". But in Città di Castello for all the people this is the "pink house", written with lowercase letters of the common things. It should also be said that "pink house" is a misleading name just as "Villa Rosa", considering that now if you want to be bureaucratically precise you must designate that building as "Villa Igea". Villa Igea is the former mental hospital in Città di Castello. The color gives to this completely trivial building a touch of individuality: it is pink. A soft pink, welcoming as a fairy tale.

I ring the bell and wait. I move slightly to take a look at the pink house placed a short distance away. At the entrance I see two overweight old gentlemen sitting on either side of the vestibule entrance, completely absorbed in listening to a music, it seems coming from the 50s. It's hard not to feel an instinctive sympathy, and while smiling I think to myself I am in the right place. A woman over forty years old opens me the door. She is blonde, curly and well groomed. I explain her the reason that led me there: I should interview a nurse who works in that structure and therefore has dealt with "insane" ("crazy") people for a long time. She asks me why, I say it is a research for the University of Florence, but that is something totally unofficial, a very informal interview. She becomes stiff. She inquires once more why, I reiterate the reason. I smile and I repeat that is a totally informal interview. This time I bet everything on that " totaly" but do not think I have convinced her, on the contrary it seems I have put her in a more uncomfortable position. She babbles something about privacy, I say again that it is nothing official but only an interview comparable to a story that she could tell to her mom or to her husband when she returns home; it only concerns her personal impressions acquired during the course of her work, and nothing else. She gets out of the embarrassing situation blaming it on a good victim: the manager. She takes me to him and tells me that indeed these are things one has to see with him, he cannot only dealing with pleasant things like giving orders around, isn't that so? I smell a faint taste of domestic disputes. With the manager restarts a deja-vu. He asks me what I want, I explain it to him. He gets a bit embarrassed. He sighs, mumbles something that has to do with privacy. I am beginning to believe that "privacy" is the most used, the most fashionable, the most chic, common English noun in Italy, of course to be used separately from any referential context. At least in that situation. I repeat the same things told to the blonde woman before, I do not know if she was a nurse or the caretaker. This time the manager relaxes, asks me anything specific regarding the interview that I want to make and thinks about who could be just right for me. Maybe now I have set my request in a better way, this improvement in such a short period of time tastes like a miracle. Or, maybe, I have to admit that I am not so next to a professional looking and therefore it doesn't take much to understand that I am not some kind of a special undercover/insider for an investigation on behalf of a "governmental" or "journalistic" agency, investigation prolific of who knows what implications. I admit, however, this thing was beginning to give me a slight pleasure. The manager is very kind, we think together who could be the interviewee. We examine three or four hypotheses and then decide he is the most suitable one. He is Gaudio, a nurse who has been working in the field for about fifty years. Actually, as I later found out, the years are between thirty and forty, and the manager is not the manager but the staff in charge. Instead regarding Gaudio we had not been mistaken, he is a competent person with a name that at a first glance interprets perfectly the person (in Italian "gaudio" means "someone gay, happy, cheerful"), and what does not mind at all: he is always willing to help. I speak with Gaudio only for two minutes, we arrange an appointment: Thursday of the following week suits both of us. I greet everybody and then I leave.

When the day of the interview arrives, it is Gaudio himself that opens the door. He does not remember that we had fixed an interview, he checks his calendar and notices a comment: "UF at 15". He starts laughing and says it was all morning since he was trying to decipher what he meant by that UF, he says he thought those letters were the initials of someone who would have to see. I actually think to myself what cabalistic formula he used to describe our meeting by "UF". He understands it well before me, and while smiling says: University of Florence.

He takes me into a room a bit small, a little hot, but after all it is difficult to pretend a cool room for August 13, and then all the better considering that my tonsils have never liked air conditioning. Gaudio is a little man, not too tall, between fifty and sixty years old, bald on the nape, with gray hair a bit disheveled at the temples. It has a spontaneous smile that lights up at the same time with the eyes. We sit, him beyond a white desk, me on the other side of the same desk. "So," he says, "where do we want to begin from?" I confess candidly all my ignorance on the subject, I would like him to tell me about his work, about how it has changed with time, things like that. In my mind the positions are clear: he speaks and I listen. For him the positions are less clear at the beginning, but then he starts talking and a trickle becomes a raging river that I struggle to stem. Just the sight of the iPod that I use as a recorder stops him a little at the beginning. I tell him that I register only because I would not be able to write everything at the moment, whereas at home I could then (successively) transcribe the interview quietly. So he keeps talking but the inflection becomes slightly more "Italian" and the pose is more professional. From time to time it is me the one who tries to be a little more "villager" ("countryman") in order to return to a more informal atmosphere which in my expectations would have been more prolific of information useful to me. Gaudio was for me a mine of information and food for thought. He commences by speaking about the first experiences of Trieste and Bergamo, then tells me that "you can say that even Perugia enters into this circle." He speaks in detail of how is organized the " madhouse " in Perugia and tells me it has also two separate branches out of the headquarters, one in Città di Castello and another one in Spoleto. I do not know at all what he is talking about, but I listen greedy of information to figure out as soon as possible where I should stop him and bring him back to his work experience at the pink house. He does not give me the cue/rise and describes in detail the "center", the Institute of Perugia, specifying well where it is and how you can reach it. Street names add to dislocating offices and in the end he starts talking about the ex-mental hospital of Città di Castello:

"... It was located in square Largo Amedeo Corsi, there at the Graces Center [Centro le Grazie], there where now is the underground car park. Before that yard was all fenced huh, and there was the wall. Then they moved it, later they put it here at Villa Igea, then there at the station, then there was a house over there at the lower square [in Città di Castello there are two main squares, their informal names are: the upper one and the lower one], but there they were alone, every now and then a nurse went there, of course, but they were alone. Then over there on top of the Gorgone, before there was a house that a private citizen had given to them, and they stayed there, and then one can say it returned home, here at Villa Igea. But before it was there, at the Amedeo Corsi square. There was a wall and there was something like a city. Thither all the patients stayed there... as well as in Perugia ... it was like a city apart, there were those who did some activities and those who did other activities. There was a barber, there was that one, there was that other one, like the military."

"I have not been in the military service, since I made the civil service ..." [In Italy in my time you could choose between the military service or the civil service, but before you were forced to do the military service; and now this is no more compulsory]

"Oh well, the military service was different. There it was like the military service, we ate in there, there was a barber, a city apart indeed."

"But was it there where now is the nursery of the sacro cuore? [literally sacro cuore means "sacred heart", here it indicates the nursery of a religious order of nuns] I'm asking because I went there when I was a child."

"Yep, right there"

"But how long ago did they do that parking lot, because, if I think of it now, I don't remember to have seen that wall as a child."

"Oh well, you're young, it might have been towards the '74, '75. That one, later, they had closed it"

"Oh well, I was born in '78, not too long after ..." I smile.

"Well, yeah, but that was before '78, maybe it was '74-'75. Before they kept them all inside, they stayed there. Then, later, they changed everything ... "

Here Gaudio begins another long monologue that keeps me glued to the chair with the elbows on the desk and the recorder in my hand leaning forward, trying to put it as close as possible to Gaudio, without making this too obvious. His initial stiffness made me more cautious. Here he tells me many interesting things. Sometimes while he speaks I would like to ask him a question about what he is saying before he changes subject. His talk is a river so in flood that I cannot stop it, so I take a quick short note to ask something when his speech would have come to an end. The peculiar thing is that within that monologue, sooner or later he always inserts responses not requested by my demand. He tells me that since the '74-'75 political leaders acted with foresight in closing mental hospitals and in trying to bring patients in their own homes. He emphasizes the rightness of those local policies. Sure, he points out, patients were never abandoned, the nursing staff watched. In addition they tried to meet their families' needs in various ways, including financially. The advantages of this new approach are obvious to Gaudio, and he also gives a practical example that I know personally. Città di Castello is not New York, who lives there, knows their characters.

"Without giving names ... But you know that big one that is always in the square? That the big man ... "

"But you mean the square of the bus?" [It is another square, it is the final stop of the city buses.]

"Yes, that one. How is he? Does he look dangerous to you? Oh consider that when he was held there in Amedeo Corsi square, he was one of the most violent. If you passed close to him, a punch, slap you took it for sure.. Eh eh if you passed near him you took his lumps."

"But you're talking about the one in the bus square, who always carries a case? Well, because I remember that when I was in the junior high school and I took the bus to return back from school and we passed by at home, there was a group of boys and girls to whom he always showed his pens. He had a suitcase full of pens, hasn't he? But that one, you mean?"

"Eh, it is unbelievable huh. And you consider that in the institute he was amongst the most violent. Oh he never killed anybody, keep it in mind it huh, but of course if you passed near him a punch from him you got it for sure. That is to understand each other, to make it clear for both of us. But as I said before ... don't think in these institutes there were specialized people, they took people from the fields to stay there. It was like the military, one was recruited according to his sturdiness. And of course they were all big men, big and strong, good-natured man huh, but strong. Medicines weren't used at all, during those times there was the straitjacket. Coercion was used and to work there you had to be strong.

Well you had to have the physical prerogatives to work there. On a place where you use force, what you want? Eh these people [the patients] became violent, the place was violent and so they became violent. Isn't it? Eh, you absorb what there is, isn't it? And that one who is now always in the square, before was amongst the most violent. There wasn't the specialized staff at all. When I took the course, it was a specialist course, organized by the Region, and they taught us how to deal with people suffering from various symptoms. But this was not done before, good-natured people, they weren't prepared at all. They used force, and clearly the patients became violent. Eh that was the environment. Then when these centers [institutes] have been closed, the patients were back to their families. But they ain't left alone. What is important is that they take their dignity back as persons. We were helping them in all directions. Consider that at the time there wasn't the card to sign in at all, nowadays if you enter five minutes before or if you go out five minutes before they ask you what you did? Why? Before you worked that's it When I was in Gubbio I had no schedule at all, there we worked hard, but we worked like it means to work huh. For me it's been ten years of fun, not work, just fun huh. What then ... But before was different as approach as well. We were prepared. We took a special course at the Region [the institution of the Region] who taught us to have an approach with the patient, that dealing with patients is important. Today these young nurses arrive, because they send them here, and when a patient arrives they treat him as if he had appendicitis. But it ain't appendicitis at all, they [the patients] need something else, you have to know how to relate with them. With the time even those young people begin to know how to relate but in the beginning ... "

"So therefore you see many differences in the relationship with patients between before and now? Do you know why?"

"Eh! Another world ... but because we were prepared, now these young nurses come, for them it's like being in a clinic; or doctor's office, then you know, now they do an internship of then fifteen days in one department then fifteen days in another department, and so they do not develop professionalism [competence]. We had done a special course that put us in the conditions of interacting with the patients in proper manner. Then later when you go around to the houses to bring the medicines, to do checkups, you need to get in touch with the families too huh. Well there you have to do a bit of everything. You have to be able to make yourself loved a little. There the doctor is multi-purpose (polyvalent), you need to do a bit of everything, and me too huh. Oh you have to know how to do things, you ain't just a nurse in a hospital, not at all. You know at the beginning when the patients were sent home you had to try to meet the families' needs, and there you had to do a bit of everything. When you go around to the house of these people you try to help them, then, you know, to the patients were given small grants, you know, trifles too, 10-15˙000 lire [the "lire" was the old Italian currency, in 2001, more or less, 2˙000 lire were converted in 1 euro, at the time Gaudio is referring to, the value was a little more than nowadays, maybe the double], at least they could buy cigarettes without burdening the family. We delivered the drugs, clearly, the families hadn't go to take drugs, we took it for them, we delivered them at home huh. After that you have to be careful because otherwise they'll ask, but you must never forget your professionalism or otherwise you lose bargaining power. Oh those families were poor too, then they asked you a hand for the bills, they wept over certain expenses, huh, and they had expenditures, after a while if you can you help them, you try to get them some other grant, another help here, you know, two cents [literary is two "soldi", the meaning is nuance, it indicates just a little amount of money] over there, a little here and a little there, hither and thither, but you do not go too far or else you lose bargaining power. You have to respect your professionalism and you have to give what is right, you mustn't go too far with the families. You must keep your bargaining power. Then we tried to get some job for the patients, something they could do, even there when there was the center there in the Amedeo Corsi square certain patients went out, they went to do their things. Then, later, there is also the personal sympathy, right? There was one that went always there, at the intersection of S. Antonio [St. Anthony] street, you know there, where now there is that bar? Well before there was a clothing store, but a great one huh. You think, it had the windows below too, was all a shop. Well that one [the patient] went there, afterwards the owner had taken him in sympathy so then had him do something, every now and then he washed the windows, stuff like that. And they gave him some coins, right? Eh everyone had something to do. Then we tried to get them some work, we went to look for some jobs at Nardi [a local industry of agricultural machinery], in other places. A little here and a little there we were trying to make them work, to have them regain possession of their person, their rights. Even when they stayed there in a house, there in the lower square, there stayed only the men, the women were here at Villa Igea, I mean there they lived alone."

"Alone? But how many were they?"

"Eh they lived alone, right? They stayed in this house, huh ... the men five or six. Yes, but the nursing staff came to check on them eh, but they lived amongst them alone, males were there in the lower square and the women were here at Villa Igea, and they were seven or eight."

"But among those people who were alone in the lower square, there have never been any problems? That is between them some trouble did not come out?"

"No, no. The nurses checked huh, brought the medicines, supervised huh, but never any problems. Then you know, it's not they were always together, consider that each one went out, had to do his own thing, these were all people who could get out. And they were going out to do their jobs, as I said we tried to find something for everyone, so they had their own things to do. If you put them in a position to do something they do it, they are people who give. Then before here, in Città di Castello, it was different, as if there was a neighbor who lacked sugar and asked it from another neighbor, it wasn't so, it wasn't as nowadays, there was more solidarity, everyone helped each other, even in there or around anyway we helped each other. When they returned to the families, then we helped each other. It was so, if you were missing something you wondered around, if anyone had it they gave it to you.... "

His talk comes to an end and I am assailed by a swarm of questions. I leave the ones intriguing me most for the end of the interview and now I exploit the footholds he has given me at the beginning of his speech and that were lost in the continuation. At first, while he led me into the closet where we would have done the interview, speaking of his work, he told me "when I was in Gubbio" within a long speech about the various types of departments that were in the "center" of Perugia. Then every now and then, here and there, always reappeared that "when I was in Gubbio", indicating that it was a very significant experience in his (Gaudio's) professional life and I believe not only in the professional life. It was good to ask him about that experience now, before the speech would have taken us to other topics.

"You first told me "when I was in Gubbio", so before coming here in Città di Castello, did you also work in Gubbio? Could you tell me something about your career?"

"How I got this job?"

"Yes, how you started, where you worked, a bit of you and this job."

"Look, I am not ashamed to say. Look, I didn't have a vocation for this work huh. There was the course of the Region [a course organized by the Region to be intended as an administrative organization], I was out of work, they paid well,

I'm not ashamed to say, it's not that I started to do this job as a vocation. Later I did the tests and then they made the competition [contest], and I passed it, I'm not ashamed to say it. It's not that we were persons with connections, but on the other hand if they require five positions and the Region makes the course and makes it for a certain number and then you fall within ... Well I ain't ashamed at all. Later I started in Gubbio, in the district of Gubbio. It's 40-50,000 inhabitants, and some places are hardly reachable huh. Successively, after Gubbio I got closer and I came here to Città di Castello. And here includes Castello [short name to indicate Città di Castello], above includes Fighille, Citerna and then up to Umbertide, Lisciano Niccone, and this makes 70-80.000 people. It 's great, then some places are difficult to reach."

"Have you found differences between Città di Castello and Gubbio? I mean socially, I mean the reception and the way to interact with the patients?"

"Well, yep enough. You know in Gubbio when they fight they fight huh. There it's a more enclosed space, and also people are more closed. From Gubbio where are you going? It ain't so many ways out, and the streets are what they are, they are closed over there. And people become closed. They are more closed than us. But look, there, if you fight you fight huh. There were also more political differences, there ... It was like Umbertide ... Well, there, the Communist Party takes 70%. And then they argue about politics. Every man wants to put in charge his own people. You see, also with the running of the candles? [It is a Christian feast from the Medieval times, but more probably coming from the pre Christian times, in which the clou is the running of the three district of the city with a big candles each. It is really particular, just consider that they run but it is not a race...] And there is Ubaldo, X, Y, together they fight badly. They begin to make some noise six months before huh. But see, if there was to give a helping hand, if help was needed, every quarrel was not worth at all. When there was a need everyone helped everyone, no matter what, there they really helped each other. Here in our places, in Castello, there is less solidarity, in Gubbio there was a lot more solidarity."

"So you found differences when patients were sent home to their families."

"Oh, there they helped more each other, were more supportive."

"Look, may I ask you if you have had a more particular relationship with certain patients? Someone with whom you have bonded more, some scene that has remained more in your mind."

"Look, here, in this work, there ain't a person to whom you bind more or bind less. You get attached to them all. Then look, they are people who give, as you work with them. They give you satisfaction. [As he says so, he is not looking at me, but gazes ahead and who knows what he is seeing.] I give you an example, if you leave the house with someone or you want to take him at the bar, and you go out from the house and maybe you have to go left, right? Here, for example, you have to go in the lower square and going there means you have to go left, but if he goes to the right, you cannot bring him to the left, you have to go with him, you have to try to bring him to the left, but without you leading him up. Maybe you're walking a longer route, but then he gets to that bar. It is longer but he reaches there, and just even this crap gives you satisfaction. What then, you don't take him to the bar just to stay there to waste time just sitting there, not at all, you take him to the bar because he comes in a context where at least he can socialize. Then other people see you sat down with him and say: look at him how he is working! But the work is to succeed in making him able to socialize, I don't want to take him to the bar, he must be the one who takes me there. In my job the difficulty ain't to do things but letting others do these things. Well, if not, if you do, you know, if you do things for them, what's hard? You do it, and that's all. Eh the problem where is it? You do it yourself, and what it takes? But it must be him to do them [the things]. He has to learn to socialize. Look, then I'll tell you this. I mention no names huh, however, it's when I was in Gubbio but not exactly in Gubbio, this one was from another town [municipality], but without giving names. I had this guy, this one is [note the present verb on some point of the story] a very intelligent chap, he graduated from high school huh [literally he got the "diploma". In Italy the "diploma" is something different from the degree, a title you get before the university degree at the end of the high school, normally at the age of nineteenth], and with the highest grades. He lived in a poor family context, poor in every meaning. It was culturally poor. The context was economically poor, but it was culturally too. And you had to face these situations of cultural poor contexts. Eh these ones were culturally poor. But he was a good guy, but really good, he was very clever: at school he graduated with honors [with the highest scores], in football he was good, in pool he was good. At a certain point he had problems in relating to other people, so then he started to say he had a number of symptoms, and once he felt pain here and once there. And his parents sent him to medical specialists, and they fell into a lot of debts huh. Consider that at that time to do a TAC [it's the CAT, it means Computerised Axial Tomography] it costed 600˙000 lire, that ain't peanuts. And he said that it hurt him there, and his parents made all the possible, but then ... they were indebted huh. And they send him to the specialists to see what he had. Consider this, when he left for the military service I took him to Orvieto, then I returned home. Oh! They call me as soon as I get home and tell me that he escaped. He got an anxiety attack. I left at four in the morning to take him to Orvieto, and nothing, they tell me so, I immediately go to his house. There good thing that I convinced him to go back to the barracks, you know after there are problems. It's not a joke. He comes back and stays there a few days, then you see, they send him to the psychiatrist. Look at the coincidence, that psychiatrist was one of the specialists to whom he was going before. He was good [the psychiatrist], he told him: I know you, you can go. He sent him home but without giving a heavy reason. He went back at home. After this I was trying to make him socialize because he is not stupid, he was smart. Look, when I took him to the bar I had to be knowledgeable [well-informed], I had to read everything, because he was always kept up to date, he always read the newspaper, then if he told me if I had saw this article here, huh, then I ought to tell him: oh yes, 1 have seen it, but if you read this other article too. You had to see that, but you had to see also beyond that. You had to be always one step ahead, but to be one step ahead of him I had to read a lot. If you are a bit ahead, you push him forward, you give him a motivation. Then you see. As I said before, here I could do that because I didn't work in Città di Castello, in Gubbio not so many people knew me so there you could work better. Here if you go out at the bar with a patient, then they tell you: look at him how he is working! But you go at the bar to make him interact, otherwise they shut themselves in, instead they can and must be with people , like everyone else. Afterwards he was good at billiards. He was at the bar and after, in short, at the bar, what can you do, there you always socialize with the dregs. With these social outcast people. He had socialized with them. With bad lots [with the good for nothing persons], with the marginalized, and so, you know, these people always socialize with these good for nothing. But these ones were rogues. Oh! Once he told me that one of them had taken his girlfriend to force her into prostitution but this girl didn't want to go and so he beat her badly. Things like that ... better pass over them ... that ... then there's this thing [he points to the recorder] ... all right... well in short he hung out with them. Then one day, just to describe what kind of fellow he was, he liked the sales assistant ... then he went to buy something, this salesgirl told him something like: 1˙500 lire. He pointed his finger to her face and in no time at all he told her: "When do you finish your shift? Later I'll wait for you". Eh this girl: " fuck you". [He laughs heartily.] She retreated backwards and "fuck you" [we both laugh], you'll understand, he was upset by that, and from there he hold back a bit. But he was doing some little chores with the Mountain Community [The "Mountain Community" is an Italian governmental authority, dealing with the development and valorization of mountain zones. In Italy they like to have a lot of Governmental Authorities]. He was intelligent, he read, he always kept himself updated, and then later he started going out with a girl of that group he knew. And you'll understand in her house he was treated like a little king. You will understand they were culturally poor people, and he was smart. Then he began to make the office gofer for the mayor or for the deputy mayor, don't remember, of that municipality there, and they had taken him in the Mountain Community, then he worked there. And then this girl got pregnant, it was a baby-girl. You see, with him, after they had moved me here in Città di Castello ... but we stayed in touch. It's not like we were in touch every day, but once every three-four months, with one of those long phone calls lasting a couple of hours, and he told me everything. Afterwards he started working at the hospital, nothing special, but for him to work at the hospital ... later when they sent him back to the Mountain Community ... it's not he was upset, he continued to do his job. And his wife got pregnant, a baby girl. But she was sick ... no, no! It was a boy, yes the second one was a boy. Now these kids are adults, they have their lives, I know they have their own thing. His father raised them eh! He raised them up, huh. And they have their normal stuff. But she got the flu, had taken a medicine for it, and then the baby was born with a malformation of the ear."

"But she had taken a medicine? I mean she had not consulted with the doctor or the doctor gave her some medicine?"

"Well she didn't know, right? She took the medicines and in that moment, you see, at 99%, she got pregnant and then the baby ended up with a malformation of the ear. The doctor didn't know, as well as her, she didn't even know, very probably the two things went together, right? She got pregnant while was taking the medicines. How could she know? Oh, look! This fellow when his child was 5-6 years, he went and took him to the specialists, but look that it would have given trouble even to me doing things like he did, to leave and to go here and there to these specialists, that poor boy did not have the auricle at all. There was nothing to be done. One day another colleague of mine who was treating him, calls me and says: Gaudio I must give you a bad news, this one had a heart attack. He had had a heart attack and died the same night, and the first thing he said to his wife was to tell it to Gaudio. Think about it, I had never seen his wife, I saw her after the funeral, and you think, he told her: tell Gaudio. After I met her, there at the funeral, but before I had never seen her at all. And who knew her! He talked to me about her, but I had not ever seen her. But this woman when she called, this one the first thing she said was: tell Gaudio. Do you understand? And who knows how many times he had spoken to her about me. Get it? But do you have any idea how many stories I can say? Here in Castle there is one that I had in my care, now, without giving names, he is being treated above, up there on top of the Gorgone, all right. Mmm what a shame ... If I knew I could have called you and we could have spoken right there. Last Friday we did something like a party, right here in Villa Rosa [the first time he calls it this way], sometimes we try to organize this kind of parties, we call the association, and we do as we try to organize these parties. Right! We could have gone there, so you could have seen and known people. I should have told you. We organize things like that, but always they pass in silence. One doesn't even know these things. Not at all."

"Well, as a matter of fact I knew nothing about it, otherwise I would have come, but are these parties open to anyone?"

"Eh! We organize them together with the associations, so then there are people and they can know other persons too, they socialize. But there is never a sounding board for these things. I could have told you, imagine there how this would have come! [He means the interview] Anyway, in short I had called this guy too. I call him on the phone, tell him that there's a party here, but this man tells me that he won't come, and I do not insist. You know, insisting isn't good. Later, the day after I went up to him, I didn't tell him again about the invitation, you see insisting ain't good, but I went to see him then if he had changed his mind he would have told me, and he was there on this institution [establishment], he stays there, and I say: "So what's up here? Come on, I find you well!" And he says: "I'm bad, here ain't people like you. I don't feel well at all. You ain't 'like them, you fill me the water." Do you understand?

They don't express themselves as we do, they are unable to express themselves, but they make themselves understood. How to say "you fill me the water" as "you give me something vital", if you don't get it you put something more than vital. He meant that I added him something vital when he needed it."

I like Gaudio, but I already start with my tendency that brings me to have sympathy for these people "mentally ill" and consequently for their "friends". I cannot do anything about it. [We think we objectively consider the data, but when we consider them we already have something in our mind that puts us in a certain direction. To know this does not mean to give up (to surrender) to this perspective.] This perhaps brings me to have "impressions" that cannot be defined as objective. But on the other hand it is natural for me to think so: where our mental construct leads us to have a starting view completely neutral? As already Kuhn and Hanson have shown, even the physical (physics) theories that appeal to criteria of absolute objectivity, based on experimental data, are not free of any starting theoretical corruption. I think to say what our starting thoughts are, is not only a duty but also a right.

One point that struck me during the conversation with Gaudio has been a peep out of the topic "politics" in his speeches more than I would have expected. When he spoke to me about the differences between Città di Castello and Gubbio, among the topics of major diatribe, he involved the political contests within the same political faction, at least in this way I think to interpret his words. Often he had also mentioned to me between the lines, even during talks outside of the interview, that, however, every elected person brought his "clients" within the hospital. This was also mentioned explicitly about his patient living in a municipality near Gubbio, who, since he was taken to make the "lackeys" of the mayor or of the deputy mayor, had been assigned to work right inside the hospital, and that, even with humble tasks, was a source of great pride for him. Sometimes Gaudio has referred to the contest through which he had obtained the tenure saying "I ain't ashamed to say", "it isn't that we were persons with connections". A not required justification often leads to thinking in the opposite direction, but this, whether it was or was not, is not the least of my interest. The point is he likes to emphasize that, and the reason could also be connected to the gossips or jokes heard and endured, which on a long term weary everyone. The politics got into the speech, and it was to be expected here, both at the beginning, when he spoke about the political leaders who started, with foresight in his opinion, the closure of mental hospitals, and at the end of the interview when he stated that it is the policy that decides which areas of medicine must be financed and developed (as you will see in the end of the interview). In his talks politics enter, politics enter neither in a surreptitiously way nor in a inappropriately way, however, it gets in and it is a fairly constant presence. Sometimes we also see how his work has been little understood by those who knew him, and that has weighed on him. When he says that his job was easier in Gubbio than in Citta di Castello, it seems to me that we can infer it does not result so much from the bigger quantity of solidarity existing in Gubbio, as stated by Gaudio. He previously points it out , but he had also asserted that in the same " before version" of Città di Castello, in the past there was a lot more solidarity than now, and this second assertion was told to me before I asked him about the differences between the two areas of Gubbio and Città di Castello. Both assertions (i.e. the more solidarity that pervaded Gubbio as well as the "before version" of Citta di Castello, when compared both of them with the nowadays/current Città di Castello) can be true , or both are to be referred to the legendary golden age that pervades a large number of people every time they turn back to review or to comment the past. The matter of fact , beyond speculation, however, is that in Gaudio's point of view the previous (before, preceding) Città di Castello was more supportive than the current Città di Castello, and Gubbio, the city where he worked before, was more supportive than Città di Castello, the city where he works now. It may be an idealized vision of the past, or, I repeat, he may be right when he says and motivates those claims, but one thing is "true" for sure and he told it to us explicitly: in Città di Castello if he brings his patients to the nearby bar trying to get them to socialize, then he is labeled by the people who know him as one who does nothing, the so-called government employee who steals the salary from our taxes. While, to the opposite, in Gubbio he had no such problem; in Gubbio, where no one knew him, there were better conditions for his job. I think these jokes, while he was at the bar with his patients, have not been so rare considering the accent with which are stressed by Gaudio, neither I think that he was immune to them.

When Gaudio talks about his patients there is a deep respect. He emanates it , and he points it out. He never calls them "crazy", even if he is aware of the jokes and gags, "then one can always make them", but he never plays one joke to me even though I asked him if they joked between them about their condition or made jokes in such way. He got the assist to make a few jokes, but he did not take advantage of it. Instead, he put much emphasis on the party that from time to time they organize at Villa Igea, where with the help of some associations they seek to create an appropriate environment to making patients socialize. His regret is sincere, and even when we said goodbye at the end of the interview, he went back on the celebrations organized at Villa Igea, telling me next time he would have let me know in time so I could participate. He never tells me the names of his patients, not even of the one he took as example of a normal life. About him I know in broad terms all his life, but I know neither his name nor his own town. I really like when Gaudio focuses on a point for him essential, that is his job does not concern in doing things, but in "trying to get them (the things) done by them (the patients)." Gaudio always emphasizes that "they", if put in a position where they can do something, develop high professionalism. He has previously said it in a light way, and will tell it, soon at the end of the interview, quoting bright and clear examples (from his point of view). His satisfaction is great when talks about his patients who, in spite of their great initial difficulties, have proven gradually to be able to live and be able to deal with a "normal" life. When he speaks about his work, about this "trying to get them to do things", he always pulls back on the seatback as if to say, it is obvious! It is obvious but it is (also) so difficult to make it clear to the others. To the "normal" people. In addition, an important role in the performance of his work is played by the relationship that he needs to build with the patients' families. That "You have to be able to make yourself loved a little", that "you have to do a bit of everything" in relation to building a relationship of mutual trust with the patient's family. This is not a secondary part of the work, and it is here that lies the difference with a nurse who works permanently in a single institute or hospital. He stresses it and holds it in high esteem. All this is part of the ultimate goal which is to try to give his patients the means to make them build by themselves a normal life. It was a fortunate fact, but perhaps the natural outfall of our chat too, to return to speak at the end of the interview about the relationship of his patients with the "normal" society, that is with that outside world of the mental health institutions. From that analysis Gaudio began a speech full of breaks, landed in a "but in the end we all are", you will understand the allusion later. I am an ignorant of the subject both from a clinical diagnosis point of view and from any other point of view relating to this working environment. But in my ignorance I always thought that normality is to be considered a Chimera fruit of the statistics, modern science par excellence. And that this normality is the product of the social context, of the specific social context. Maybe in the future, to give an example, an autistic person will be regarded as the model of a man most suitable and functional to that (future) new world to come. The normal man is the result of the average between individuals different from each other, who are not normal by definition but considered such because are functional and useful in that context.

When I asked him what was the correct name to call these institutions ex-asylums, I thought that in his answer I could have grasped a certain his preference or a few comments, or even more certain ideas to be explored; instead he addressed the issue pragmatically, I would say bureaucratically. I do not think that he did not leak any preference for an abundance of caution, but simply because for him the name is totally irrelevant.

"Well, these were called mental hygiene centers (in Italian: centro di igiene mentale), then, later, they were called mental health center (in italian: centro di salute mentale)."

"But with the change of name also changed a different approach to the patient, or was it a pure label change?"

"Who knows ... No ... I'd say differences, not at all, it's always been the same only at a certain point they have changed the name from mental hygiene center to mental health center. Dunno (to translate in English the "famous" Italian expression "boh", it is quite hard, anyway it means something like "I don't know"), I don't know. Maybe with health is less offensive. Boh ... What do I know? Maybe it's a more neutral term."

The final comment of the interview was dictated to me by him, nothing else to be added. In fact I ask him if between them they joke about the name usually assigned to them by the outside world: that is "crazy" ("crazy people"). If sometimes it happens to him to joke with his patients about this reputation that is attributed to them as a category. I refer also to Gubbio, which, at least in Umbria, is famous not only for beauty but also for being the town of mad people. (If you go there you can get also an "official" certificate, just following few instructions...) The same pink house is also called the "House of Crazy/Mad". I ask him if he jokes with his patients of "pink house" on their status as "crazy", if between them they joke about that. His answer had nothing to do specifically with neither the "pink house" nor with Gubbio. For him the answer is not related to the single workplaces nor to the various contingencies.

"Actually ... the problem is how we approach [our attitude], actually you see that one is someone a little crazy, so always prevails having a little bit of pity, that part of being afraid to appear, of vanity, actually he is a person who has equal dignity, if we put ourselves in a relationship of equal dignity, I think, isn't it? It's the most correct thing, it's the most correct relationship, then, you see, one says: "He comes from the House of Crazy", it's true, here there is everything, right now as I speak it is less posh, during the 70s the psychiatry was fashionable, I say during the '80s began to be popular the toxic addiction, so clearly they are choices, right? At first the psychiatric discourse has been funded and encouraged, then services like SerT and things like that [SerT are public services for the toxic addiction, they give psychological, diagnosis, orientation and therapeutic support to the toxic dependent and to his family], then there was the AIDS problem and then the infectious diseases that were a consequence, let's not forget that the hospital of Castello [shorter name of Città di Castello], that new one, when it was designed, an ad hoc department for infectious diseases was created that was supposed to serve for these diseases, then because the opening was delayed and the new opening was in 2000, the infectious diseases went already off fashion, because in the years after 2000 was fashionable the so called dementia, Alzheimer's, and so obviously even the choices at the local level, at the political level, the economic choices changed, you change what you finance …. You finance better such services than you fund others. But this is like a trend in the entire Italy, huh. Then in the discourse of crazy I sometimes start to laugh, like … I laugh because I can … right? Then I remember that this one that was a boy, I remember when another one took a gun and fired on the main street [of the city center] … "

"Where here in Castello?"

"No. He shot a signboard he had seen. That one has become the head of the election office of the City, he ain't from Castello, I won't tell you from which municipality, right? But it was a municipality near Gualdo [another city of Umbria] and he was himself for the elections that ran all the bureaucratic machine, the entire election from the ballot to the poll and after, it was him, huh. Bank managers, not to mention local doctors, of Perugia, then after maybe you find someone of them in some places … and the normal people say: you know I've been to the doctor … even because in their professionalism it's people who are good and the furthermore ... well, as I am saying, did you close it [alluding to the recorder]? "

"No, it's still on."

"Well, by the way as the relationship with that guy before, I took ... in the meanwhile he was still spending time with the rabble, I took three days off, as I was back to work a colleague of mine says to me: "Look, it's three days morning and afternoon that he always comes to look for you." I see him, I told him: "And what's up?" He's like: "Look". He pulls out a revolver with a serial number "imae" [a registered gun]. This gang gave it to him because he was "the best" [he smiles, and the tone of his voice with him]. I remember I took him upstairs and I gave him a really bad roasted, so bad I can assure you someone else would have gotten pissed off. He went away as a hangdog. He went to Umbertide [another small town in Umbria, near Città di Castello] and threw away the revolver into Tevere [the river Tiber]. And then he came back. To say that you reached the maximum. Yeah, but it was a maximum, get it?"

"But in your opinion, when you said before that anyway, the " scum ", those people low down in the society, are always the most open about these things. However, it is true, the bar ... "

"Well, indeed! It's so."

"But do you have an explanation for it? Because anyway it's true, when I was younger who joked more with them [the "crazy" people], those one who had a little more ... "

"Well, probably ..."

"I don't know, maybe they have less hesitation."

"Less hesitation certainly, much less hesitation. But here those ones more available are, we can say, the most marginal ones. The reason for that I never ... honestly here we are, I never wondered why. Perhaps, one finds himself really with less rules with less hesitation less modesty [shame], also he would show himself how he is. Because that would be a lot too, isn't it?"

"Yes of course."

"Eh we all got that [that shame], at least the concern to show how one is actually. This sometimes puts in serious difficulties eh, particularly these fellows, these patients, in short."

"I see. Thank you, you have been really a source of information."

"Did you have enough? Ain't that a lot?"

I would say no, actually it is not enough for me. But I abused far too much of Gaudio's time, I asked him for half an hour, and instead we have already exceeded it by an hour. In this case I would rather not follow the Latin phrase *melius abundare quam deficere* ("it is better to exceed ,more than the necessary, rather than run the risk of not having it in a sufficient measure, rather than to be at shortage") and so today I decide opt for the "deficere " (for the "shortages"), in order to have an open door just in case I would be in need in the future.

I thank Gaudio and ask him if, in case I would need some clarifications on what he has just told me, I could still bother him. He answers with a smile: "Of course!" Actually I have other things in mind: I would like to interview another nurse too of the same area (field), possibly of the same " Villa Igea", but one that has a different cultural background from Gaudio, in order to have a different perspective on the same topics. I exploit the week during which Gaudio is on holiday, I do not think he would take offence for that, in case he does I would gladly explain the reasons, and I am going to ring at Villa Igea in search of a certain Cristina, a nurse just employed there.

FAILED INTERVIEWS

I know Cristina indirectly, she is a friend of a friend of mine and peer (to her), between twenty and thirty years old more or less, she has been working for three years at Villa Igea and this fits perfectly with the profile of nurse who I would like to interview to. When Gaudio talked about the surgery-oriented preparation of the new nurses, it made an impact on me. For this I ask a friend of mine, who is also a nurse but in the hospital, if she knows someone working as a nurse at Villa Igea, and if she could ask for her/his availability for an interview. My friend tells me about Cristina and fixes me an appointment with her once more at Villa Igea, once more during the working hours. I come provided with the just recharged iPod and a small notebook with a few notes and annotations to ask. For my astonishment things will be very different from how I had envisioned them. As soon as I arrive, Cristina herself opens the door asking me, totally embarrassed, what would be the interview all about. While blushing she says: "What would you ask?" I tell her what every reader of this paper by now knows very well: "Nothing special, something concerning your experience here, your relationship with your patients, your impressions. Whatever you want. It's an interview for an examination at the University of Florence. It's totally informal and unofficial." She is really embarrassed, she is becoming even more embarrassed, as if it was her turn to do something that she just did not want to do, but did not know how to avoid either. She tells me she would not even know what to say, she speaks about the well-known privacy. To reassure her I tell her that I already did an interview with a colleague of hers, Gaudio, but successively I had a problem with the iPod, and during an automatic synchronization with another device the audio recording of the interview was deleted, so now I had only the last part of that interview because it was saved before on another hard drive. Therefore I would like to ask her just two quick questions just in case I would have needed the recording. All this is true, but it is clear that I use it as an excuse, the interview had already been transcribed and what interested me was not as much the recording of the interview as an end to itself, but rather to hear her impressions and opinions on certain issues. However, I talk about this problem occurred to me in order to explain why I want to ask these two questions and nothing more, considering she keeps asking me "why".

I would like to reassure her that this is not an investigation of the Israeli or Iranian secret services. Cristina is very embarrassed and while the blush is becoming ever more visible emerging from the heavy makeup that adorns her face, she says she would do everything possible to contact Gaudio so that he can redo the interview. In short, it was impossible to interview her; maybe because of her insecure nature, maybe for fear of saying something that would lead her to be accused of a wrong approach to her work, or perhaps for some other reason that I do not know, the fact is that seeing her so much embarrassed I do not insist further. Seeing her so worried, I empathize with her and I feel a little uncomfortable too. Smiling I tell her that she should not worry, that there are no problems at all considering that the interview to Gaudio was "ok". I reiterate that this interview would have had the only purpose to be at most a kind of an addition or corollary, and that it would have been a "safety" interview considering the loss of the previous recording. Nothing more, so nothing to worry about. Seeing her so apologetic, I tell her that there is no reason to apologize. But I cannot help to notice the similarities with Marco.

Marco is another nurse from another mental health center, the one that was indirectly referred to me by a former patient of Gaudio's, with whom Gaudio has remained in touch. Gaudio reported that this patient was not so good in that institute, where he actually lived, because nurses there are not like "him" (like Gaudio), "they do not fill me the water". Marco is rather young, he recently passed thirty years old, but he has a good experience behind, he has been working at the institute for about ten years. I agreed with him for an interview, but after the unnecessary and especially the so numerous restrictions that he imposed, as well as the embarrassment he also had to talk about his work, I preferred to change direction on another nurse to interview, so I went directly to the "pink house". Marco did not want to be registered, did not want to meet me at his place of work not even to fix an appointment for the interview to be done elsewhere, he would also have granted me the interview (if I really had no other possibilities) only in some far location during the cocktail hour, but his name wouldn't have been mentioned at all and we would have had to agree on a fictitious name that would have suited both of us. Bin Laden would fix less restrictions to a CNN reporter. (By now I could also say "would had". This paper dates back to the summer of 2010.) Anyway the names used in this paper are all fictitious, there was not even the need to arrange this. Marco has the looks of a quiet boy, less groomed compared to Cristina, who, later when I went to interview her on the job, or at least I thought I would have done so, was dressed and made up with great care and refinement. But both of them were so incredibly embarrassed, clearly this "incredibly" is to be referred to "for me".

I do not think they did not want to talk because they were hiding who knows what big truth, nor only for shyness. I think that being so young, especially Cristina, they have been afraid to talk about their work just because, unlike Gaudio, did not master yet well their working environment. It was only because of insecurity. This is a shame. I regret because Gaudio talked about his non- vocation for the profession, and I was curious to ask them about it too. They are far younger than Gaudio, coming from another cultural and economic background, I wanted to know the reason why they had undertaken this profession. Being two young nurses, in truth I had prepared some questions. I imagined that Gaudio both for his experience and because he did not even opposed to the interview, would have faced it glib and with attitude, saying things that I, ignorant of the subject, would not even ask at the beginning. His only moment of hesitation was when he saw a recorder in my hand, but it was a moment and nothing more. With Marco before and with Cristina afterwards, I knew it would have been different, but between "different" and "nothing at all" there is a big difference. Sure their non-words are not regarded as "nothing". It is obvious. I do not know if, as Gaudio stated, the young nurses have a surgery-oriented approach with patients, " as if patients had appendicitis", however they demonstrate for sure, as I interpret these non-interviews, a lot of insecurity in the performance of their work. This can also be considered normal, but I remind that Marco has been working in the ex-CIM (Italian acronym for Mental Hygiene Center, that is the former name of the current Mentaly of Health Center) for about ten years. Perhaps the long and exhaustive talk of Gaudio compared to the silences of Cristina and Marco can lead to considerations too merciless (for the last two). Nevertheless, this is the result of my brief research under the pressure of the "privacy".

I must point out that in these institutions/medical centers the appeal to the privacy follows you closely in the same manner as a hunting dog hunts the game running away from the lords riding on their horses, in their well-dressed suits, wrapped in their stylish red jackets and in their tight white rider pants, which make you sick to the testicles only at their sight. Every time one enters and asks for something informal in these environments, they answer you calling on the privacy, it might also be right. Indeed, I remove the "also" and put the verb at the present time: it is right. But considering that a person does not enter in a hospital only to ask for and to do these interviews, as any reader well knows, I want to lead you in the place and at the time when that curly boy over there is Lorenzo, and that paper he holds in his hand is a normal request from the family doctor for some blood tests, designed to detect the presence or absence of certain "diseases".

His girlfriend had had some little troubles and he did not want him to be the cause, much less Lorenzo would have wanted to be the cause in the future. This why he was there for the blood tests. The time is the past, not too distant , the place is the waiting room of Città di Castello hospital fit for the payment of the fee for tests and analysis. When it is his turn, Lorenzo approaches the nurse "cashier" and gives her the sheet showing the required analysis. Before I tell you what the nurse said I would point out one small thing: has Lorenzo mentioned what kind of analysis were there on the paper? Has he mentioned what the problem was? No. To tell the truth Lorenzo didn't speak at all, actually he did not say a word about it even to other "users" there in row with him (in a hospital whether we like it or not, we are all users before being patients). It was the nurse the one who took care of informing everyone about it. A little intimidated, Lorenzo explains to her that he had already explained the problem to the family doctor and that he had prescribed the analyzes reported on that paper, the same one that now she was holding; on that paper there were the stamp and the signature of the doctor in order to make it well distinguishable from the prescription that he could have been made by a Siberian shaman or a pantheistic blue component of the tribe of Pandora / Avatar. The result: the nurse insisted that they would speak louder, but she did not have to insist for a long time, and before Lorenzo could have reiterated his reasons, she warned the users of the last row, that:

"You can't! I know it, I'm an expert nurse, what do you think? These blood tests have nothing to do with these health checks. You can't do it, because here you need a test to be done through the rectum ..."

Aware that one swallow does not make a summer, I report an additional, true, example for the record. Do you see that worried guy over there? He is waiting for the results of an HIV test, since a doctor, going wrong in the interpretation and deciphering of some symptoms, had previously prescribed this to him. But in that present moment he still does not know that an error occurred in the diagnosis. The boy is a bit intimidated. When they call his number, he approaches the desk with an alternating stride between being firm and being hesitant. He asks for his analyzes in a low voice. The nurse does not hear and asks him to raise his voice. Otherwise, if you don't speak louder, one cannot hear anything! Then she takes the analysis, looks at them, and giving lessons in opera she says to him:

"Eh, you were lucky this time, huh. You didn't get this AIDS, come on, stop worrying. But next time you'd better wear the condom, you will enjoy it anyway!"

Not all nurses are so loyal in front of his eminence the privacy.

APOLOGY (NOT REQUIRED) OF THE NURSE

Sometimes the doctor and the nurse are accused of not giving credence to the Hippocratic Oath, sometimes they are accused of confusing it with Hypocritic. But we should not fall into the same fault expecting others to do what we are not able to do ourselves. It is pertinent here a consideration that should be noted: for his protection a normal nurse must create a barrier between himself and his patients. The patients are in medical care, maybe some of them are facing degenerative states from which they will never recover completely, some will not recover at all. The nurse who has a nature "to dive" into the patient, who has a propensity to have an almost empathic relationship with him and takes care of every patient in the same way he would look after his family member, wouldn't have any suitable salary with which to be rewarded by the society. In this context, the same would be valid for a doctor. It is the gold of medicine, and in this case medicine should not be understood "only" (!) as a science for the good (health) of the man, but as solidarity in itself, true and not vulgar. But the reality says that for the majority of men, and therefore for nurses in general, to be in contact all the days with sick people leads necessarily to develop a barrier between them and the patients. And there is nothing wrong in it. Far from it! We cannot expect each nurse to suffer for every patient that he has on care in the same way as Mother Teresa of Calcutta is said to have done in the past. Otherwise, even without dragging out mysterious miracles, each nun would be a saint. And every saint would be a nurse or a doctor. And it is not so at all. A nurse is not expected to enter in the path that leads to sanctification, nurses are not asked to cancel theirs lives for others. They are expected to be professional. For this the society pays them a monthly salary. A deserved/earned salary. After all, not everyone in their life achieved a job that allowed them to have a reward which brings a livelihood and a role in the society, just following their nature and their "heart". Gaudio admitted it. There are people who succeed in obtaining the job they always wanted to, there are people who end up by chance in something in which then they will deeply involve themselves, there are people who have had to adapt to do something for a living. Everyone will put the passion he has in his blood there where he wants, where he succeeds and where he can. However what is required is the professionalism. This is valid for every work environment and for the nurse too. Those who expect that other people give in their job much more than what themselves put into their own job, fall into that human category placed by Dante Alighieri in the sixth pit (bolgia) of the eighth circle of the Hell: the hypocrites. What we, as citizens of a community, could say at our present state of living beings belonging to our organized and institutionalized societies, is that we are compelled to claim professionalism from a nurse, as indeed in any other working environment. [Although by now that word, citizens, seems to me that should be considered subject to updating]. For the nurse who manages to put more than professionalism in his work, the reward is the respect and a heartfelt thank. Well, obviously in addition to the salary.

THE OTHER HUMAN FOCAL POINT/ FOCUS

In Italian the title of this chapter would be literally translated as "The other human pole", but like all the literary translations it would lose something from its meaning. At the opposite a less literary translation but more aimed at giving the idea, would give something in plus to the meaning. Something is taken off and something is added, it seems a law so present in every area of our lives that looks like a physics law.

Speaking of "nurses" and "crazy" as well as of "human pole" would presuppose their positioning at the antipodes of the same social system, sometimes probably it is so within that social system. But in "reality" they are two different social categories (social categories interpreted so by the society) of the same context, of the same group both social and species, etc. Therefore they are aggregation centers within the same context not necessarily at the antipodes as the word "pole" would let us presume, that is why I would prefer to use here the term "focus" (than "pole"). The Italian language perhaps is more ambiguous about it, English more precise.

The two focus points are the "nurse" (in Italian "l'infermiere") and the "crazy (man)" (in Italian "il pazzo"). For the word "nurse" there are not substantial problems in the translation. Instead for "crazy" there are plenty synonyms in English, perhaps much more than in Italian, and each one has a different nuance/hint. Perhaps referring to persons who are in care, persons that are defined patient, it would be more appropriate to use the word "insane (people)"; but this is a negative definition of a human "category", i.e. starting from another category [the "sane (people)"] and so defining itself as opposed to that category. In this case, yes, they would be "poles" located at the antipodes. But this is not the meaning of what I am looking for.

Actually in the Italian piece of writing the indecision I had, was between "matto" and "pazzo". If "mad" has phonological connections with the Italian equivalent "matto", both phonetically and etymologically speaking, there is not as much equivalence between crazy and "pazzo". The choice felt upon "crazy" and shortly we will discuss about that. Anyway it stands to reason that the two terms, one perhaps more used in England and the other one used on a wider scale in America, are basically synonyms, and in a such manner we have used them and we will continue to use them. But a choice had to be made. Aware that a name in respect to another brings to mind something different, I opted for crazy both for etymological and phonetic reasons. It is relevant the fact that both terms have phonetic echoes in other languages of Indo-European context, thereby showing a deriving from common terms/vocabulary of substratum. Supposing that a certain human Indo-European nucleus existed, with its related uniform language (the matter seems likely at the present, although not without problems), the reconstructions of the Indo-European lexemes "pazzo" (deriving from the same root of "patient")" and "mad/matto" would call back from the meanders and depths of our mind a primitive and original idea of those men. Such a situation gives to these terms/words a considerable diachronic depth, and if the signifiers were used it means for sure that there were also their referents. The madmen have always been here (existed), and just this is a comforting consideration.

In this diachronic sphere "mad" seems to have a more widespread use and therefore a better (established) deep-rooted use. "Mad" would derive from:

the old English "gamædde" meant "out of one's mind", "extremely stupid" ("usually implying also violence")(..)from Proto-Germanic *ga-maid-jan, demonstrative form of *ga-maid-az "changed (for the worse), abnormal" (cf. Old Saxon gimed "foolish", Old High German gimeit "foolish, vain, boastful", Gothic gamaips "crippled, wounded", Old Norse meiða "to hurt, maim" (Online Etymology Dictionary)

"Mad" emerged in the Middle English and it was more connected with the word "anger", even if successively that was considered only as an American use. However, as a matter of fact, nowadays "mad" is more used in America than in England and in that context it competes more with "angry".

Despite the phonological parallelism (similarity) (same initial syllable followed by a dental sound in the island of the industrial revolution and a deaf one in Italian) it does not seem to have etymological connection with the Italian "matto", however the phonetic similarities are strong. In ancient Provence, region of France, we have "matou", in Lithuanian "mattas", perhaps both to be connected with the greek word "màtaios" indicating the fool, the demented (man), the useless (man). Certainly this is not a compliment to be considered among the titles. In Gothic then we have "midus", perhaps even to be associated to ancient German "metu", a term designed to indicate something intoxicating, particularly linked to certain beverages. The German man with a beer in hand is not a modern invention. There is a reason why in the late Latin, period of strong Germanic influence, "matus" or "mattus" were used to designate the " drunk ". In addition we have the Sanskrit "madu" and in the chess game of Persian derivation (Persian is the "1.2" of the Sanskrit) we call "checkmate" the move that puts to an end the opponent's king, and with him the match too. Checkmate in English has no meaning, as well as the Italian equivalent "scacco matto", litteraly something like "the chess is mad/crazy", these terms are the phonetic adaptation of the Persian *sciach* (re) *Mate* or rather *Mat* (is dead), "the king is dead" (from "mat-", to kill). Actually in the Anglo-Saxon world it is considered an Arabic misinterpretation of the much less bloody Persian verb mandan-, litterally "to be astonished", "to remain alone". Anyway it reminds us that the goal of the game of chess is to kill the opponent's king , not to eliminate the highest possible number of opponent's pieces (the not too hidden reference is for a ten year old boy who is learning to play). To the contrary of the English "mad", more linked to "anger", the reference of Italian "matto" to a "game" it is not a coincidence. Consider an Italian derived word from "matto": "mattacchione" (from a former "mattaculone"), usually translated in English with "joker". The "mattacchione/joker" is a person that is a bit crazy, someone capable to joke, it is a playful term/word attracting something like a hilarious/cheerful meaning. To draw a conclusion in Italian a mad person is someone "dominated by irrational impulses, by uncontrolled gestures (actions), by unusual and excessive mania, however, such characteristics and mania raise more often hilarity than apprehension, sympathy rather than aversion". (Devoto-Olii) On the whole, one cannot note that actually this term "matto" would be not suitable to describe a "patient" who occasionally or on a permanent manner must be followed by specialized medical personnel.

"Crazy" is the perfect word I was looking for. This word is a bit more recent compared to "mad" as well as the Italian "matto" and "pazzo", nevertheless it covers exactly the innermost meaning I was looking for. And that's why I chose it. It is attested since the Medieval time, perhaps deriving from the Old Norse *krasa*, literally "shatter", giving a sense of "mental breakdown". Later on, at the end of XVI sec., according with the Online Etymology Dictionary we find *craze+y* that stands for "full of cracks or flaws", giving the meaning of "of unsound mind, or behaving as so". During the period in which was at his highest, in the 20s of the XX century, this term was used in the jazz slang to stand for "cool, exciting". Moreover "crazy" has something romantic, it carries itself with a romantic aura, a romantic light wind. You easily say "crazy love" rather than "mad love". "Crazy" became also a popular term when it had to be included in the name of a famous Indian warrior commonly known as "Crazy Horse". Actually there was a little misunderstanding in the translation of his name, because in the Dakota language the "crazy" was not to refer to the Indian warrior but to his horse, to be correct his proper name should be translated as "his horse is crazy". Who knows why during human history the enemy or simply the others are more misunderstood than the allies or the "friends".

Much less incomprehension has arouse from the etymology of the Italian word/term "pazzo", that here we associate to the English "crazy", neither for an etymological nor phonological point, but only for the current meaning that the two words have in their contexts. Etymologically "pazzo" is supposed to derive from the Latin *patiens,* that has also led to the Italian lexeme "paziente (literally in English: patient)", but which originally meant "suffering ", a "suffering person". This is in fact the Muratori's view, who precisely considered "pazzo" as derived from the Latin *patior* (to suffer). Some scholars propose to compare/juxtapose it to the ancient German *parzjan* or *barzjan* too, both words "soaked" with the meaning of "to infuriate ", "to enrage". Moreover the Greek has the term *pathos*, expressing both an infirmity of the body and an infirmity of the soul. The Greek tragedy arouses the pathos, and in its presence we are all equal, all human. We are all "pazzi", all "crazy". It makes us all crazy. So here we return to the "pazzo" (crazy) deriving from the Latin *patient*, "a suffering person" ("one who suffers"), in this case to associate inseparably to a state of infirmity of the mind. But sometimes dividing the two things is not so easy.

I chose "crazy", as well as "pazzo" in the Italian version, because it fits perfectly with the examples brought to us by the specialist nurse Gaudio, with the "crazy" Damasippus protagonist of the third poem of the second book of the satires of Horace, and with the person that I will introduce you shortly. Then these terms also have another feature, this time not an etymological one but a totally phonetic feature. To pronounce both words, "crazy" and "pazzo", you are forced, like a snake, to spurt out from your half-closed lips a hissing sibilant sound. Crazy is the serpent that deceived Eve and led to Adam's conviction, crazy is the snake who stole from Gilgameš the miraculous leaves of the plant of youth. These are jokes like those usually played by the joker, but this time they have nothing hilarious. At the same time the sibilant consonant sound makes us smile when we continuously listen to it coming from someone affected by zetacism. Nevertheless the word crazy is sparkling, with that final "z" that makes us a little like children and intimidates us a little like the sound of a snake.

The etymology is a fascinating science, it seems to reveal the world that unconsciously one carries inside his head. The etymological survey, so formulated between these lines, transforms our mind in an archaeological field where one proceeds to identify the various layers of meaning accumulated and sedimented under a specific sound or word during the history, and during the flowing of events and different social-ecological contexts. All that sedimented in a specific sound, in a word. However, the risk is to end up tangled in it, and playing too much with it leads us just to agree with Voltaire when he stated that "the etymology is a science that turns everything into everything". Unfortunately for eighteenth-century alchemists he clearly was referring to the words and not to matter.

A madman is for the society a particular member, and that can be seen by the attention the society itself has for him, attention which sometimes I do not know how much can be beneficial. But by interviewing him seems to me to move him into the " freaks " category. Even from the Gaudio's interview, and it is a point of view of someone having a much deeper relationship than mine with the madmen, it emerges that a "madman" talks and expresses himself with facts, not with words. The words customarily twist on their throat like ivy on an oak tree. Even the same fact of deciding to not interview a crazy man, in opposition to what I did with the nurse, places the first one in a different context from the second one. But, at the end of the day this reflects the same polarity on which this writing moves. For me, but only for my personal feeling, by not interviewing him I recognize the specificity of the crazy, instead by interviewing him I turned him into a circus freak, where I would let you in for the price of the ticket represented by the reading of this short writings. It is not what I want. I choose the first option. I prefer to let the crazy speak with the facts, facts that have to be limited by a well-defined time frame, same as the interview was, in such a manner to protect a parallel point of view on the "outside reality" to which one accesses using different channels.

I have many crazy friends, and some mad. Or so I think. Then you know, for me it is always valid the rule of "who knows who", and I follow it like a monk of some Benedictine monastery follows the rule "ora et labora". As many others I look around with curiosity, and meticulously I catalogue according to certain mental tendencies. A librarian of lived (experienced) events. I have no Eskimo friends, but I have a lot crazy friends and someone mad. Or so I think. One of them at some point in his life was under treatment and had to embark on a quite intense course of medication. I never attended a university course that would lead me to have greater expertise in this field of medicine, but from my point of view these intense pharmacological cures sometimes are nothing more than a surrogate, and also quite poor, of certain human relationships and true motivations of life of which at times, sooner or later, we all feel the need. We conceive the brain as a nice apparatus (machinery), this is not the ultimate dock of the knowledge, which I do not think exists, but only a functional view of the world surrounding us. "Every mean is a social tool" (Offner 1996). This applies a fortiori to the intellectual means. Surely when we will have completely understood the mind, when the brain will have no more secrets, we will be able to act on its "mechanisms", on the proteins, on the synapses, modifying ourselves according to our needs. Maybe then it will be important, more than now, to understand and to give a derivation, a source, to that "according to our needs". Then everything will be perfect, everything will be solved, everything will shine if we will want to make it shiny. Then we will instill optimism within the brain limbs, and we will see that everything will be solved, we will be thereabout feeling happy and satisfied without even needing it. Happy about it? Unhappy about it? Who cares, now we are not at that point. For those who are waiting for this point, for who is not at this point, for those who believe that it will come, for those who do not believe it will come, the "truth" is this: we are at the point "here and now" and we are not in front of the point just described. And waiting for its arrival in front of us, I think that where the nowadays medicine cannot reach, the philosophy can reach. The way we understand the world around us, and the way we feel immersed in it, it is the first clear source from which we draw our life. Who has an idea that leads him will always be superior to those who do not have one. And I say this as an atheist.

I spent words to explain why I chose "crazy" instead of "mad" for the title of this text. However, the example I have taken is a minotaur between "mad" and "crazy". Joker and romantic like a crazy, suffering like a madman. This could be a necrology that would be good on the grave of each one of us, and it is certainly a caption that applies to a large number of my friends.

But among these crazy friends of mine a prominent place is occupied by Luigi Remigi from Città di Castello also known as Gigi, or The Boss, or The Prophet. [The fact that he is from Città di Castello is relevant for the continuity of the framework environment with the nurses interviewed previously. And then deep down that city is one of the protagonists of this writing, and such it is shown in every page.] The boss has short blond hair, a nose slightly arched, on the model of a Pazienza's cartoon character, and he is a young boy of thirty-three years old, he will remain so even at sixty years old. This year (the thirty-third) in the life of this man has inauspicious precedents. He is tall and very robust. More than robust one could say big, and now more than big one could say inflated. He has always been robust, but he has become inflated since he was prescribed some medication that would have helped him with his "problems". He can ask you the same question seven times in eight minutes, these numbers are not random, and he reveals a stunning sensitivity. At the time I am writing, the artistic "verve" has got him again. He turns up at your house unexpectedly and after a while, he never stays for long, just a greeting of a quarter of an hour, he tells you he is going to make you a painting "because you are a great". And here he makes you choose: do you prefer that I make you a Van Gogh or a Picasso? I want to outline him for you in the same way he did the picture he gave me. It is a beautiful canvas of a flowered meadow blown by the wind behind a fence. I briefly introduce you Luigi Remigi, as a nineteenth-century impressionist would have outlined him for you.

In summer, it is not difficult to find him in certain woods and scrubs overlying in the nearby of the locality "il Sasso", intent to prune the trees of the wood. Well if he does not take care of it , "nobody does". For a while he often went to the Tiber River to collect stones along the shore, with them first he filled up the car, and then he filled up the bathtub. He said they gave him energy. But I think it was the same energy that was removed then from his mother when she had to clean up and remove the whole stuff, in fact now he has stopped doing that. You know how it is: unleashing the "dark side of the force" is risky for everyone. One day I met him in a supermarket, he had bought some spices and joking I asked him if he needed them for bathing too. He thought puzzled for a moment, then said thoughtfully: "No, I am not a chicken or a duck!" After another three seconds he burst out laughing heartily. I remember this sketch not because it remained in my mind then, but because a few months later, when I returned back to Città di Castello, I met him again and the first thing he told me while laughing was: "Oh that, the joke! It was beautiful huh. You hit me, hit me!" During a period when he was having depressive problems, he was sent in a community with the hope that it would have done him good. In the community there are people who have "serious" problems, "normal" people who have fallen into some vortex for some reason who knows how and who knows why. Each case is an unique and different, but certain answers that come to mind there should be good for many cases, and both me and you know them well. People for whom is actuated a "shock therapy", in order to save them socially. Remigi was there, because other solutions did not work, so even this one was tried, and also with him it was actuated this strong therapy. About that experience, he told me that the priest did nothing but telling him he was a slut, that he was fine and the people who suffered were others. The treatment (therapy) included a total isolation from the outside world, relatives, friends and especially parents. Conclusion with touch of class: Remigi ran away from that community close to Christmas, taking away with him Baby Jesus from Nativity scene. He was then taken back home by a doctor at whose house he accidentally had stopped to make a phone call, he got lost. While he accompanied him home with the car, the doctor also prescribed him a different treatment (therapy): a couple of visits to some prostitute.

He has always been a very religious guy, he attended, and I think still now he attends, various parish church groups in the same measure he hangs out in several pubs where they play rock and blues music. After all, another of his nom de guerre is "bluesman". He has too much vital energy, too much lymph, to be exhausted by a calm conversation in a room in the shadow of a crucifix. In his red car that seems was sewn on him for how it physically fits to him, it seems he is wearing it, not driving it, a red narrow and high Matiz, bears the symbolically Christian image of the fish. At his neck till to some time ago he wore a huge wooden crucifix. A while ago he came to visit me at home without wearing anything on his neck, I said: What happened Gigi?" "Franciscans have pissed me off!" I never knew why or if he was reconciled. This I have to ask him.

He keeps asking me the doctoral thesis which I did on the Hittite glyptic of BöğazKöy, but give it to him means to dislike him. So every time he requests it again he adds some "ethnic" conception, a new " eschatology ", some hints of his personal " epistemology " , and a strong desire to live that sounds like a request for help. He is an amazing person, as anyone who knows him knows. I cannot conclude this brief sketch of the Boss, without mentioning his supreme passion, by now too well known even to those who do not know him personally: the music and his guitar. At every party he arrives with his guitar and plays and sings. He spices up everything. At the wedding of a friend of mine, he succeeded to throw a small band out from the raised dais; a little band called, and paid, just to liven up the party. And he succeeded in it with the same nonchalance with which the moon revolves around the earth. Natural events do not need to be forced. He plays everything. He lingers from pure blues to any music for party, from U2 to Battiato. But the song that stirs the crowds is the unmistakable "La Menca", strictly performed in "Castelano" dialect (the dialect from Città di Castello). He has also written some songs of his, the first of them that I recall was proposed to the "general" public of the Alberto's bar located in San Giustino (a little city near Città di Castello), a bar that later we discovered did not belong to Alberto, he was the waiter. Just that says far and away about the turning out of those evenings, where a girl wouldn't pass by not even if you have paid her gold, and where, when for some strange astral coincidence, she would have passed by, then the same waiting silence would have fallen down like the one with which some thirty male souls would look at S. Peter when during the Final Judgment would be just looking out from the balcony and would be about to return the verdicts. The same waiting silence of the English parliament in front of W. Churchill who is about to declare war on Nazi Germany. The same silence with which I listened to a song by Luigi Remigi: "alienated". He plays it often, and he does well. It is his song, it is him. You can see through it. Last summer, 2010, there was a nice initiative in the park of the meander of the Tiber in Città di Castello. On the occasion of the falling stars night, in this park a group of guys organized a series of little shows interspersed with "celestial" visions. Now, I admit that the idea was better than its execution, and that the execution was much better than the technical means at their disposal: few and of very, very poor quality. Here Remigi could not miss, with his guitar he played his song: "alienati (alienated)". Even more beautiful was his dedication, unfortunately there was not a lot of people over there, but this suited him better. "This song is dedicated to all those who ... To those who have more ... More ... In short, those who have more... Those "more"! Come on, I know you got it! "

His other passion, but without comparison less than the music, is painting. Every friend of his has one of his paintings, as I said I have a beautiful landscape of a flowery meadow behind a fence in a "Van Gogh" style. At a certain point he gave up a bit this thing, then an afternoon on the terrace of my grandmother's house I asked him why. After all he puts so much life into it that those paintings trickle of him, like the portrait of Oscar Wilde exudes of its owner's youthfulness.

And then with the paint one pours creativity into something material and this is supremely beautiful (at least for me). The answer was:

"Eh Leo, I know. But then you know it takes time, then one has to follow up on it, then it takes inspiration. Then ... Huh ... And then they don't understand them and my mum doesn't know where to put them, and she throws them away! "

He made me die from laughing because I had already seen how many of his paintings filled in his living room, he is a serious painter, and so it seemed to me just to see his mother in front of me, that, while mumbling because of the mess, was moving the paintings and herself here and there, without being able to find a permanent position for all those "Picasso". After all, where I come from, grandmother Pina teaches: the mumbling of the women in the kitchen is as much typical as the calendar of "Soothsayer Monk" (literally "Frate Indovino", is a calendar concerning the astrological prediction of a monk) hanging near a radiator or a stove. However I do not know if Remigi's mother had or not such calendar.

His mom should be made saint for the unconditional love she has for her son. And for her patience too! For his thirtieth birthday Luigi Remigi took us all to eat in a country tavern (trattoria)-pizzeria called in secret code "dal duro (literally: "from the though man")", or even as "lo zozzo (literally: "the filthy")", where you pay little and you eat. The attributes are not used randomly. Now it is much better, but at the time Luigi liked it more. There with his guitar he could start an endless big mess. And it was the only restaurant where they were also happy that this happened. I still remember how he made laughing an English family who came down for dinner from who knows what agritourism in the neighborhood. In short, for his birthday it was him to bring a gift for every invited: all things actually lived. Useful things and not ends in themselves, supposing that such things exist. And another thing in common for all that stuff, it was they all came from, but all indeed, his house. He gave me a kitchen utensil, that I still have, useful, I think, to cut small cakes or more probably to pour sugar. To each one of us a gift from his kitchen, and on each of the gifts there was a dedication for everyone. He is a great man. We all appreciated it, his mother I think a lot less. Instead my mother appreciated less the fact that in an afternoon while I was with her in Garibaldi's square, I met Luigi and I stood there talking for about five minutes. We talked as if we were facing one another, and indeed we were, but staying in opposite ends of the square. My mother gives a lot of importance to the bon ton, I had not even realized that we were yelling, attracting the attention of every passer-by of the open air market. It was on a Thursday. Elected-day for "grandmothers" 's open air market, on the contrary the one on Saturday is "open" even to the young girls.

I am sorry in these pages I cannot reproduce and I cannot let you hear his voice, his tone. He speaks in a low and husky tone, pushing out words that do not want to go out, that would like to stay with him, but then come out in a sigh. When he gets excited, and this is not a rarity, his voice bursts powerfully, high, strong, you can hear it from a great distance. He is basically a very shy boy, he is the first to know that, and the second ones who know that are the girls with whom he spoke. Like all those made this way, he tends to show himself in the opposite way, especially when he was younger. With his husky tone that gets even higher, stronger, and at the same time almost incomprehensible. But when, as usual, he talks with his tone a little low and a little husky, with those words pushed out gently as if they were given off by a thirty-three laps not well maintained record player, then, that is his voice. You can hear it only if you are near hm, but you can hear it pretty well. Thinking about what Remigi says is not possible without having on your mind the sound of his words. That noise that echoes like the swashing of the waves on a deserted beach.

As I said I am not going to interview Luigi Remigi. He is my friend not someone to know. His interview will be transcribed here with the story of the last time I saw him, that is the night of New Year's Eve 2010. I do not know if anyone will read this writing, not even when it will be read, but for me, that I'm writing now, the new year of 2010 means yesterday. Well, clearly successively I reviewed what I wrote.

We had a dinner between friends, the Boss played the guitar for most of the time, both upon request and without, obviously, livening up the evening, and then, as a good tradition, we all went to the main square to celebrate. At about 2:30 he asks me if I see him to his car, and he asks me that in a way (that) no one could say no. At least I couldn't. While I am having fun with my friends in the middle of the square, I feel a tap on my right shoulder, I turn around, I see the Boss making me a Mona Lisa smile and says: "Leo is already late, what are you still doing here? We had fun, we go home, come with me?" The low tone with which he asks me that, dominates each phoneme used. In theory to get to his car would have taken about fifteen minute walk, so I consider also the possibility to come back to celebrate a little more after having gone with him. And then in the case I did not return, I would not mind it, in any case my girlfriend was on the other side of the world. From the main route we took three different "panoramic" diversions, we reached the car after an hour and a half. So we had a chat and a long walk too. Many speeches that begin and end, but in the middle point there is no half. Useless to try to reconstruct them in a exact and faithful way, but they were fabulous. All speeches focused on the spaces of Città di Castello. This time I had no IPod to register, but certain things are registered by themselves. "For us men everything is fine, everything is great for us humans, but for the people then it becomes tight." He speaks to me about lime trees along the city walls. "They are not taken care off badly, but they could be kept better." Gigi had a bottle of water that bothered as a car parked sideways in the middle of a cycle lane, a quote from a poet. Any person would have given the bottle to the first passer-by who needed it, or he would have thrown it away; Luigi Remigi waited for the right tree to water. He offends the driver of a passing car who looks and pays attention to other cars but not to pedestrians. We are the pedestrians. As we pass a dog starts barking. "Oh stop it or you will get a bronchial pneumonia." Shortly after he is like: "Some dogs are dumb as some of their owners." We pass right in front of Villa Rosa, and shortly after a "certificated crazy man" walking by himself, asks us for something. I do not understand what he wants and I say blindly "no thanks". At the opposite the Boss gets it, he stops in front of him and says "An euro no, but I have a cigarette. Wait, I'll look for it." He puts his chubby hands into the forward pockets of the pants, then into the back pockets, then he takes them off after having searched well, then puts them into the pockets of the jacket, he keeps searching in such location for another fifteen seconds, maybe more plausibly another twenty, and finally, with his chin pushed inside, with a touch of embarrassment that turns his face in red, and with his low and husky voice, says "I'm sorry but I don't have it." This fellow gets close to Remigi, they exchange two kisses on the cheek which on that night are worth as wishes, and then they resume their walk each one in their own section of the road. After a while, he says:

"This fellow was crazy for real!"

Remigi is like clods of earth just plowed in the autumn fields of the tifernate rural area (tifernate means from Città di Castello). Like those large, dark, brown, turned clods from where come out earthworms and various insects. A swarm of life in a dark intense brown background. A clod among many others, more moist and waterlogged than many others. It makes me think to that madman who in the interview to Gaudio said "they don't fill me the water", to be understood as a symbol of life. Remigi is full of water. I see him passing among those clods, for those same ploughed fields of the Tiber valley and lingering for hours watching a farmhouse-rustic house, absorbing every memory that emanated from the ruins.

Remigi is full of life. One night we were at the graduation party of a friend of ours, we went out together from the building to get some fresh air for a moment and there he spoke quietly to me about his suicidal tendencies. About many other things. At one point he said, with his voice hoarser and marked by a great slowness, as if those words had to travel a long way before coming out from his mouth: "Leo I don't give a fuck. I'd rather kill myself, but if my mum would be so sorry ... Oh. Well ...Then I do not do it." You cannot read this sentence in the same manner you can read a Mc Donald's slogan. "The phrase" he told me during my friend's graduation party. When he tells me that phrase, I seem to see the perfect samurai who just came out from the Hagakure. Certain words arranged in a certain order and expressed in a certain way, smell both of life and death at the same time. The words might not ever express the essence of the life, but they let you get a hint of it! That moment reminds me each intense moment of my life, and all of them together. The final of the 2006 World Cup in Florence in San Niccolò, taut for the football match and taut for a not too timid blonde American girl seated in the grass near me; a summer night in good company ended up in a swimming pool on the high set of a mountain; the incessant chirping of cicadas during the sun-drenched summer afternoons; an afternoon when as a child I saw my younger and chubby sister jumping and bouncing happily on a big trampoline while a small group of boys did not spare themselves in uncalled stupid jokes until my father came to take her away and to offend them, slightly toning down the anger whipping in my body; the same effect with which I am absorbed in the big smiling eyes of Georgiana Gherghina; the same value of the last drop of sweat that falls before you exhausted finish the race; the same ecstasy of a child who from the bottom till the top watches the defender of humanity, the robot Mazzinga Z, found in a Christmas Eve's night not far from the fireplace, etc. etc. Remigi with all the serenity of a pachyderm grazing the grass told me: "Leo I don't give a fuck. I'd rather kill myself, but if my mum would be so sorry ... Oh. Well.. Then I do not do it". I will never be able to render the idea of the man who spoke to me in such terms about life. Seneca from on high of his Stoic awareness could not have done better before soaking in the bathtub for the last time. His words were not the result of the resignation. They were the result of the serene contemplation of a forthcoming summer night. In a place that I do not remember neither as bad nor as beautiful, Remigi was dissecting the being, the life, and this not because he did not know it, but because on the contrary he was loaded of it. Loaded is the right word. I do not know which one of you has ever been in the countryside during the early morning of the summer season, when the morning dew is not too cold and it stands out distinctly as water and not as ice. It is a phenomenon not so special, it can be perceived also in the center of Milan, or in Bucharest, or in Hyde Park, there only has to be grass there. But there, in the countryside, it is stronger; there is that primitive lifeblood that flows stronger in a tree deep-rooted in the ground and concentrated only to survive, to live. It is there, during the morning hours, that you see stems of grass overburdened with dew, bent by that water, but that you would never describe in those terms. You see them. And you describe them as rich of dew. Rich and not overburdened. Life in Remigi is like the morning dew on the summer grass stem. Remigi is full of life, Remigi is overburdened by life. Only those who are rich can be overburdened.

McCarthy has written a book, "The Road". It is about a man and a child, a father and his son, alone, desolate, in a dead world, where there is nothing, not even light. A few survivors who strive to survive and suppress one another for their own survival. Throughout the book, the man and the child are hungry, anesthetized by the stench that they themselves give off, they get worked up to eat something that barely separates their skin from the bones. They are sick. Emaciated. They stink. Who reads it considers this book a painful one, distressing. Many have read this book, and many have called it distressing, depressing. It is a hymn to life. Two lonely people, a man and a child, hungry, exhausted, sick, lonely, in a ugly world, but that in spite of everything , and without knowing why, they want to stay alive. They want to live. Not because they are surrounded by luxury, by the perspective of winning everything they touch. But they want to live just to live. Which hymn to life could be bigger? I do not know if there is something bigger, but I think Remigi is the same hymn to life. Remigi is full of dew. No billionaire will be able to suck the sap of the life of who is entirely absorbed only by his survival. And nobody would swap the position of the former with that of the latter.

BIBLIOGRAPHY

AA. VV. a cura di: Albera, D., Blok, A., Bomberger, C., 2007, *Antropologia del Mediterraneo*, Milano.

Augè, M., Herzlich, C., (a cura di), 1986, *Il senso del male. Antropologia, storia, sociologia della malattia*, Milano.

Breda, N., 2000, *I respiri della palude*, Roma.

Boncinelli, E., 2006, *L'anima della tecnica*, Milano.

Devoto, G., Oli, G.C., 1994, *Il dizionario della lingua Italiana*, Firenze.

Durkheim, E., 1962, *La divisione del lavoro sociale*, Milano (ed.or. 1893).

Fabietti, U., 1998, *L'identità etnica. Storia e critica di un concetto equivoco*, Roma.

Fabietti, U., Malighetti, R., Matera, V., 2002, *Dal tribale al globale. Introduzione all'antropologia*, Milano.

Fasulo, A., 2009, Fiducia, *Antropologia Museale n. 22*, pp. 51-53, Roma.

Goffman, E., 1988, *Il rituale dell'interazione,* Bologna (ed.or. 1967).

Habsbown, E. J., 2006, *Il secolo breve (1914-1991),* Milano (ed. or. 1994)

Harvey, D., 2010, *La crisi della modernità,* Milano (ed. or. 1990)

Ladyman, J., 2009, *Filosofia della Scienza,* Roma (ed.or. 2002).

Luhmann, N., 2002, *La Fiducia,* Bologna (ed.or. 1968).

McCarthy, C., 2007, *La Strada,* Torino (ed. or. 2006)

McLeod, C., 2008, *Trust,* The Stanford Encyclopedia of Philosophy (ed.or. 2002 Massachusets Institute of Technology).

Kuhn, T.S., 1969, *La struttura delle rivoluzioni scientifiche,* Torino (ed.or. 1962).

Nietsche, F., 1978, *La nascita della tragedia,* Torino (ed.or. 1871).

Offner, J.M., 1996, Gli effetti strutturali dei trasporti: mito politico, mistificazione scientifica, in: *Geografia delle comunicazioni. Reti e strutture territoriali,* pp. 53-67, Torino.

Piasere, L., Solinas, P.G., 1998, *Le culture della parentela e l'esogamia perfetta,* Roma.

Schrippa, P., 2009, Sistemi Medici, *Antropologia Museale n. 22,* pp.129-131, Roma.

Serrano, M., (a cura di) 1995, P.C. Tacito, *Germania,* Milano.

Tylor, E., 1970, *Cultura primitiva*, Bari (ed.or. 1871).

Zadeh, L.A., 1965, Fuzzy sets, *Information and control n. 8,* pp. 338-353.

-----   , 1968, Fuzzy algorithms, *Information and control n.5, pp. 94-102.*

About the Author

Leonardo Massi (Città di Castello – PG, 1978), historian, hittitologist, geographer, anthropologist. After two degrees, one in Antique History (Hititology) and one in Geographical and Anthropological Studies (Geographical Information Systems), a Ph. D. in " Civilization of the Antique World" and academic publications in some of the most prestigious magazines related to the history of the Ancient Near East, presently teaches Economic and Touristic Geography at the "Istituti Tecnici Superiori" of Treviso.

leonardomassi@ymail.com

www.ingramcontent.com/pod-product-compliance
Lightning Source LLC
Chambersburg PA
CBHW070425180526
45158CB00017B/752